TRACING NEW ORBITS

COLUMBIA STUDIES IN BUSINESS,
GOVERNMENT, AND SOCIETY
Eli M. Noam, General Editor

COLUMBIA STUDIES IN BUSINESS, GOVERNMENT, AND SOCIETY

Eli M. Noam, General Editor

Tracing New Orbits
COOPERATION AND COMPETITION IN GLOBAL SATELLITE DEVELOPMENT

Edited by
DONNA A. DEMAC

COLUMBIA UNIVERSITY PRESS
New York   1986

Columbia University Press
New York Guildford, Surrey
Copyright © 1986 Columbia University Press
All rights reserved

Printed in the United States of America

Library of Congress Cataloging-in-Publication Data

Tracing new orbits.

/ (Columbia studies in business, government, and society)
Bibliography: p.
Includes index.
1. Artificial satellites in telecommunications—
Congresses. I. Demac, Donna A. II. Series.
HE9719.T73 1986   384.5'1   86-9593
ISBN 0-231-06344-X

This book is Smyth-sewn.
Book design by J. S. Robert

Contents

CONTENTS

Acknowledgments

I wish to thank those individuals who have contributed most to my launching into the exciting orbit of satellite communication.

Many of the people working on space-related research, including myself, have more than a little of the whimsy. In visions as well as research, I received incomparable support from Philip Mattera.

Professor Eli Noam, director of the Research Program in Telecommunications and Information Policy at the Columbia University Graduate School of Business also merits my appreciation. It was he who invited me to plan and organize a conference on international satellite programs from which many of the papers in this volume are drawn. Roberta Tasley, executive director of the Columbia Research Program, has frequently rendered me much valuable assistance and was a creative spark in the formative stages of the satellite conference. Special thanks also go to Anne McKay, a telecommunications consultant, for her help with this project.

For several years, I was a member of the Federal Communications Commission's Advisory Committee that worked on U.S. preparations for the 1985 World Administrative Radio Conference of the International Telecommunication Union. The committee chair, Steve Doyle, made this a rewarding experience by keeping the floor open to a spectrum of ideas. Loring Chase and

Heather Hudson furthered my understanding of the issues before the I.T.U. and helped me "stay the course" by their friendship during the best, worst and most circuitous of meetings.

George Codding, Jr., Ram Jakhu, Heather Hudson, and I prepared a background report on the I.T.U. Space Conference in the winter of 1985. This was a timely opportunity for me to dwell upon some of the themes covered in this book. The project team made this a productive and stimulating project and I thank them individually.

At New York University's Interactive Telecommunications Program, it was Professor Mitchell Moss, then Department Chair but normally based at the Graduate School of Public Administration, who first asked me to teach a course on the regulation of satellites. I will always be grateful for his decision to have me develop a curriculum on a subject which was already of great interest to me. Also at NYU, I thank Red Burns, the current Department Chair, and my students over several years who have spurred my work in the area of telecommunications.

Taking eighteen papers and making them suitable for publication takes many hours of close attention to text. I did this job with the able help of Kate Wittenberg, my editor at Columbia University Press and Beth Silverman, who assisted with copy-editing and wordprocessing. Nelly Trillon, Kathleen Gygi and Rebecca Lentz provided valuable assistance in the latter stages of this project.

Introduction

DONNA A. DEMAC

In 1985, the space shuttle Discovery went into space with five Americans, a Saudi Arabian prince, and a Frenchman on board. During the weeklong journey, two communications satellites were launched for the Mexican government and a consortium of twenty-one Arab nations, while laser and biomedical experiments were conducted for the U.S. military and several countries in Europe. These activities were paving the way for the next era in outer space. The principal objectives of the first era, global connectivity and technological breakthroughs for both the space and earth segment, had been accomplished. The new objectives include the use of space stations for scientific and industrial applications, and the establishment of worldwide, multi-user, multi-purpose information networks that will include satellites, digital switches, glass fiber cables, and mobile communications.

Like the space shuttle crew, communication has become truly international. Over 1.5 billion television viewers in sixty-eight countries watched the Los Angeles Olympics in 1984. One hundred and ten countries are members of INTELSAT, the core network for telephone transmission worldwide. A similar international organization, Intersputnik, provides satellite services to the Eastern bloc and some developing countries.

The proliferation of satellites amidst the introduction

of sophisticated telecommunications technologies is transforming business activities and national perspectives on international communication. There are today approximately one hundred and sixty commercial satellites in orbit. In the industrialized world, satellites are being used to provide telephone service, data communication, video, and specialized networks. In the developing countries, India, Indonesia, Brazil, and Mexico own satellites which are used for business, education, entertainment, and national unification.

Satellites are now of recognized importance to the economic progress of most nations. The ability of satellites to establish telephone connections quickly facilitates national communications and export. Satellites are now being used by large multinational corporations as well as by smaller enterprises to accelerate access to information and perform business transactions.

The costs of building, launching, and maintaining satellites are enormous. For this reason, bilateral and multilateral participation have turned out to be advantageous at every stage, from manufacture to launching. Increasingly, the satellite universe is comprised of hybrid combinations which have both cooperative and competitive characteristics. These include global and regional satellite systems, public and private satellite ventures, multiple purpose satellites and multiple frequency systems. In several regions, organizations such as the European Telecommunications Satellite Organization (Eutelsat) and the Arabian Telecommunications Satellite Organization (Arabsat) have been formed to permit the pooling of resources and minimization of political tensions.

Satellites are relay systems that defy national boundaries. From signals transmitted at a distance of 23,300 miles above the earth (the geostationary orbit, where nearly all communications satellites are placed to permit continuous transmission), countries in Latin America are likely to receive television signals intended only for the United States. The satellite footprint marches boldly into neighboring countries, oblivious to whether it has been paid for or is wanted by all those whose land it encompasses. The political frictions that can be generated by such situations are considerable. Countries are trading charges, on the one side, of "piracy," and, on the other, of "cultural invasion."

Fundamental policy issues have been raised in connection with the rapid expansion of international satellite activity. Much of what is taking place is in response to markets that provide opportunities and incentives for the development of new technologies. The economies of scale require international marketing of equipment and services. This puts pressure on national regulatory agencies to determine how they can be effective in the overall context of a technology that is international on various levels.

Not all countries will opt for the "open skies" policy of the United States. Open markets mean more competition, more complexity, and diminished control. Like all telecommunications industries, satellite progress depends as much on national traditions as on the technology itself. Countries are fashioning different regulatory strategies based upon the existing regulatory structures, national customs, and resources.

The earliest ambitions for satellites focused on global connectivity. This gave rise to membership satellite organizations, "common user organizations," most notably, INTELSAT. Most countries cannot afford and have no need for independently owned satellites. Leasing or sharing alternatives are the most practical means to improve domestic telecommunications.

At present, however, the one-network approach that led to INTELSAT is being reexamined. Countries in Western Europe and throughout the developing world are planning regional satellite systems. In many cases, they are interested in obtaining transmission capacity in addition to that which INTELSAT provides, yet it remains to be seen how new international satellite transmission can be introduced without endangering the existing distributive equities of the INTELSAT network.

The challenges to INTELSAT provide insight into the options for international cooperation and competition in the future. Several papers in this book look at this situation. Some background is in order here.

In March 1983, the Orion Satellite Corporation filed an application with the Federal Communications Commission to establish a two-satellite communication system over the North

Atlantic (a heavy traffic route for INTELSAT). The company's plan was to sell transponder capacity on these satellites to television networks and other large corporations to meet their internal communication requirements. Orion argued that it was pursuing the economies of specialization while protecting INTELSAT where it should be protected.

The most important factor in this situation is Article 14 of the INTELSAT agreement which provides that public international satellite systems established by INTELSAT members outside of the INTELSAT framework shall avoid "significant economic harm" to INTELSAT. As INTELSAT derives much of its revenue from the North Atlantic region, the application by Orion and the four companies which followed called attention to what this competition might do to the global network.

Apart from these U.S. companies, there are other plans by governments and corporations that also are likely to affect INTELSAT's predominance in international satellite transmission. Some nations are seeking greater national independence in transnational communications and hope to build domestic industries that could prosper through the construction or use of new satellite systems.

In other parts of the world, which until now did not need a large amount of satellite capacity, there is room for additional capacity. The general direction is toward more decentralized satellite communication systems, which means that decisions have to be made as to the proper way to introduce competition and decentralization. The next developments in international satellite activity will be exciting and yet demanding. A new frontier is being settled. How satellites may enhance, rather than override, existing forms of communication is one important consideration.

Entrepreneurs often behave as if history begins today. Yet even in the young satellite field, this is not so. Twenty years ago, the earliest explorers of outer space did something unique in human history: they declared outer space to be a commonly held resource. The Outer Space Treaty of 1967 is a statement of the basic ideals and principles of international space exploration, including the principle that outer space is not subject to national (or private sector) appropriation. It is to remain open to the use of all nations who are expected to share the benefits of space development on an equitable basis.

This treaty was adopted before space became as important as it is today for military planners and commercial enterprises. It means that the enlightened visions of the early explorers are today in jeopardy: how can they be carried into a more complex environment, one with many more nations, interests, and possibilities?

The papers contained in this volume examine many of the issues that must be addressed in order to bring about stable, peaceful, and equitable satellite communications in the future. In the overall context of cooperation and competition, these papers analyze important policy issues, point to areas in need of research, and offer specific recommendations.

The book is divided into five parts. In the first, "International Satellite Activity: Challenges of the 1980s," a paper by Carl Christol, noted legal scholar on outer space law, reviews the most important international space treaties, and calls for deeper examination of the common heritage principle for commercial space activities on the one hand and for the expanded militarization of space on the other.

Fiber optics is finding its place in the telecommunication revolution. Data are translated into light pulses that travel through glass fibers, are amplified, then converted back to analog signals. Integrated broadband communication networks of the future will rely on fiber optics, but also on satellites. Roy Layton, president of Sears Communications, examines the competition between satellites and fiber optics from a technological and cost standpoint, and even finds ways in which they may complement one another.

The four papers in part 2, "The Emergence of Specialized and Regional Satellite Systems," look at the present challenges to INTELSAT. Marcellus Snow provides important historical information and analyzes the economic arguments for and against private satellite networks. Joseph Pelton, director of strategic policy at INTELSAT, argues that the core network should be preserved and that consultation should be a requirement in creating alternatives or add-ons to INTELSAT. Christopher Vizas of Orion Satellite Organization, the first company to apply to the Federal Communications Commission to establish a private satellite condominium in the North Atlantic, describes Orion's outlook and predicts that many countries and companies will soon be offering services in addition to those that INTELSAT provides. Douglas Goldschmidt, an economist with experience with satellite projects in the Third World, presents some of the issues for developing nations on the INTELSAT question.

The third part of the volume, "National Programs and Perspectives," includes five papers about satellite developments in Western Europe and North America. A paper by Brenda Maddox, communications editor of *The Economist* magazine of London, compares the approach of the United States in favor of deregulation and maximum competition to the more cautious and regulatory approach of governments in Western Europe. Mario Hirsch, an executive of the Coronet Satellite Corporation—the "mouse that roared" from Luxembourg in 1982—finds fault with the slow approach in Western Europe and regards the weakening of national television regulations as inevitable.

Turning to Canada, W. M. Evans of the Canadian Department of Communications, describes what is essentially a hybrid government/industry strategy toward satellites. Canada had the first domestic communication satellite and, today, has a highly developed satellite program oriented toward expanding communication domestically and exporting satellite technology and expertise.

The United States program, prodded by the deregulation of telecommunications in recent years, is moving toward increased commercialization. A paper by Fred W. Weingarten and Charles Wilk, policy analysts at the U.S. Office of Technology Assessment, discusses the implications of the federal government's research and development in telecommunications for the satellite field. Jerry Freibaum of NASA then describes his agency's agenda for the remainder of the decade, explaining the U.S. approach to commercialization. Sylvia Ospina, a lawyer who recently studied at the Institute of Air and Space Law at McGill University provides a stimulating paper on satellite signal piracy in the Western Hemisphere.

The fourth part, "Regulation of International Satellite Activity," contains four papers by experienced observers and participants at the International Telecommunication Union, a UN agency with responsibility for regulating worldwide use of the radio spectrum. This is an important time for the ITU. The first part of a historic conference at which new procedures will be adopted for allocating slots in the geostationary orbit was held in 1985. The ITU's responsibility for bringing about equitable access to the orbit is now receiving the closest consideration in the union's history.

All the papers in this volume were written before the 1985 conference. Each writer comments on the central issue of the 1985 conference—equitable access. This issue will not be decided by the ITU until the second half of the conference in 1988. Thus, these papers are still highly relevant to present developments.

The first paper in this section, by William Montgomery, presents the concerns of Canada in connection with the current ITU discussions. Then, Heather Hudson, who has spent more than ten years as a consultant on the use of telecommunications for economic development, concentrates on the issue of equitable access. Hudson shows that meeting this objective inevitably will require attention to such issues as training, low-cost technology, financing, and organizational procedures.

William Dizard, of Georgetown University's Center for Strategic and International Studies, argues that the importance of INTELSAT, Intersputnik, and other multi-user systems (common user organizations) in making satellite communications more accessible to countries merits more attention in the ITU's discussions of equitable access. ITU membership is available only to nation states. Because of concerns about sovereignty, the importance of communications organizations has often been played down.

Finally, Harvey Levin, an economist at Hofstra University, calls for closer attention to the anxieties of developing nations. He calls for empirical studies that will provide the information necessary to evaluate the fears that cost factors and un-evenly paced satellite programs could prevent poorer nations from having equitable access in the future.

A volume on the topic of cooperation and competition would be remiss if it failed to look at satellite communications on the part of the Soviet Union. The last section, "Soviet Satellite Communications," looks at the satellite program of the Soviet Union and at the possibility that satellites will allow people of the East and West to know more about one another. A description of Intersputnik is provided by John Downing, chairman of the department of communications at Hunter College. Then, an exciting innovation, an earth station that brings in Soviet programs at Columbia University, is described by its inventor, Ken Shaffer. Shaffer describes ways that satellites can be used to build new channels of communication, bypassing the walls of ideology and suspicion.

The papers in this volume are a contribution to the long-term challenge of making the most efficient and equitable use of satellite technologies. In the coming years, we may expect to see more collaboration among competitors and more competition in the context of joint ventures. The articulation of national and business objectives should help to forge stable links among nations in space. Considerable sensitivity and policy acumen will be needed on the part of all nations.

TRACING NEW ORBITS

PART I: INTERNATIONAL SATELLITE ACTIVITY: CHALLENGES OF THE 1980s

O N E

The Search for a Stable Regulatory Framework

CARL Q. CHRISTOL

A principal function of the international legal process is the formulation of generally acceptable international legal principles, standards, and rules. These prescriptions contain rights and duties which apply to the conduct of states, international intergovernmental organizations, and other juridical and natural persons. When a state enters into an international agreement it becomes obliged to adopt laws designed to secure the objectives specified in the agreement. In this manner the juridical and natural persons of States are required, pursuant to municipal law, to conform to the international norm. They also acquire rights allowing them to engage in activities of their choice.

It is within the foregoing framework that international space activities will be carried out. Since limitations exist respecting activities in space, it is important that the process for establishing such limitations, and the critical elements of international space law, be identified. Having looked at these matters it will be possible to respond to the question whether a stable regulatory framework exists. One can then proceed to an examination of what ought to happen in order to improve on the present regulatory system. In embarking on such an inquiry it must be borne in mind that mankind will increasingly make demands for new space activities.

EXISTING INTERNATIONAL NORMS

Formal international norms relating to the exploration, exploitation, and use of the space environment (outer space, per se, the moon, and other celestial bodies), and the natural resources of the foregoing areas, have resulted from the efforts of the United Nations and the International Telecommunication Union. In the former the role of the Committee on the Peaceful Uses of Outer Space (COPUOS) has been supplemented by the efforts of the Conference on Disarmament (CD). In the latter the periodic meetings of the ITU have been supplemented by international and regional World Administrative Radio Conferences. Five formal agreements of major importance have resulted from the deliberations at COPUOS. The constitution of the ITU, most recently revised in 1982, and the regulations adopted in WARC sessions have also conditioned space activities.

These formal agreements, as augmented by an expanding body of general rules of customary international law, based on accepted common practices of the space-resource states, constitute today's regulatory framework for space activities. This legal structure constitutes an acceptable substantive basis for international satellite activity. However, the legal framework for space activities is incomplete in some important particulars.

In order for the international legal regime for outer space and space activities to be complete, particularly as the commercialization of the space environment and its resources move forward into its next and enlarged phase, there will be a critical need for the establishment of a regulatory institution. The exact nature and function of such an institution will be the subject of much debate, particularly whether it should be charged with the management of all space activities, or given only a limited mandate, while preserving, for example, the present separate rule of the ITU. Such a new international intergovernmental organization at a minumum should be given the constitutional authority to secure the implementation of the UN-negotiated treaties. Such an institution should also be empowered to promulgate and to apply rules resulting from its own deliberative processes.

Foremost among the existing principles of international space law are: (1) the space environment and its resources are to be used "for the benefit and in the interests of all countries . . . and shall be the province of all mankind";[1] (2) the area and its resources are to be open to free scientific investigation; (3) while states and international intergovernmental organizations may exercise jurisdiction over space activities, neither may establish sovereignty or equivalent authority with respect to such activities; (4) space activities are to conform to international law, including the UN Charter; (5) nongovernmental entities may engage in space activities, subject to the international responsibility of the parent state; (6) liability shall devolve on the parent state for damage caused by space activities; (7) countries may call for consultations in order to resolve concerns respecting the safe and efficient use of the area and its resources; (8) on the basis of reciprocity States may visit national space objects and the facilities of other States located on the moon and other celestial bodies; (9) astronauts, as envoys of mankind, are to be given help when in distress.

The foregoing principles, which have their source in the 1967 Treaty on Principles Governing the Activity of States in the Exploration and Use of Outer Space, including the Moon and Other Celestial Bodies (Principles Treaty), must be read in connection with a limited arms control provision appearing in Article 4 of the agreement. In this article the parties have agreed not to place in orbit around the earth nuclear weapons or other weapons of mass destruction. The agreement does not prohibit the orbiting of other types of weapons. Further, Article 4 provides that the moon and other celestial bodies, but not outer space per se, are to be used exclusively for peaceful purposes.

The 1967 Principles Treaty contains basic principles allowing for the development of scientific and commercial uses of the area and its resources. The treaty does not contain provisions designed to limit the kinds of weapons, other than nuclear and mass destruction weapons, which may be introduced into orbit. It is also deficient in that it does not require that outer space per se be used exclusively for peaceful purposes.

Subsequent treaties have clarified and extended some of the provisions of the Principles Treaty. The 1968 Agreement on the Rescue of Astronauts, the Return of Astronauts, and the Return of Objects Launched into Outer Space requires that assistance be provided for the recovery of endangered astronauts and that they be returned safely to the launching country. It also calls for the recovery of space craft that have made an unprogrammed reentry, and, on request, a return to the launching authority.

The 1972 Convention on International Liability for Damage Caused by Space Objects imposes a duty on a launching State to pay for damages to objects on the ground, to aircraft in flight, or to persons or property in orbit. The agreement imposes liability on national participants and on international intergovernmental organizations for harm resulting from their joint international space activities. It ennunciates a formula to be used in fixing the monetary sums to be paid in the event of harm. It also makes provision for the resolution of disputes over damages, including the establishment of a claims commission.

The 1975 Convention on Registration of Objects Launched into Outer Space requires parties to register their launches on a national roster and then to give notice of the launch to the UN Secretary-General. Such notice is to include the date and location of the launch, orbital parameters, and the general function of the spacecraft. Notification must also be given when the space object is no longer in orbit.

FACTORS INFLUENCING COMMERCIAL
SATELLITE ACTIVITY

Profitable exploration, exploitation, and use of the space environment and its natural resources will depend on a myriad of influences. Some are presently known; others can only be imagined.

International Considerations

The existence of a stable international regulatory framework for satellite activities will depend on a number of political considerations. At a minumum there will have to be an accommodation between the competing views of the space-resource States and those comprising the developing world, i.e., the less-developed countries (LDCs). The issue here will be the equitable sharing of the benefits derived from the area and its resources. There will also be the need for a balanced approach between the free-enterprise system of the western world and the economic-political formulas preferred in the socialist countries. Somewhat ironically the commercialization of the area and its resources has gone forward concurrently with the evolving militarization of the space environment. In this area the United States and the Soviet Union will be obliged to arrive at positions based on mutual self-interest in order to allow for the maximization of commercial undertakings.

The excessive militarization of the space environment will constitute a hazard to important commercial developments. Governments will be required to provide economic support, at least in the beginning, for private commercial activities. Funds budgeted for military purposes may diminish the sums available for the support of commercial activities. A large number of military launches will burden launching facilities and could preempt radio frequencies required for future space stations, which can be visualized as a convoy of orbiting space objects surrounding a large central facility.

Commercialization will have to take into account frequently expressed Soviet views. Although the Soviet Union agreed to the terms of Article 6 of the 1967 Principles Treaty, which provided that "nongovernmental entities" may engage in space activities, Soviet spokesmen as recently as the October 1984 meeting of the International Institute of Space law stated that they were not "happy" about the prospects of private commercial activity. This outlook may be based on the existing bias in favor of state-owned enterprises. It may be founded in the view that a large number of commercial launches would result in the occupation of orbital positions and the use of radio frequencies, which would deny that area to competing launches. It could be that the Soviets have the view that such private launches might serve the intelligence-gathering needs of the launching states, with possible injury to Soviet security requirements. It may also be that the Soviets consider that such commercial launches would solidify the existing relation between earth-based and space-based activities with the possible result that the early users of orbital positions could lay claim to preferences.

Commercialization of the space environment and its natural resources will increase space activity. Commercialization will take many forms. At the present satellites have been successful in augmenting communications. They have been engaged in a variety of sensing or monitoring activities. Agricultural and fisheries yields have been improved. Critically needed weather information has been provided.

The experiments conducted on the space shuttle have measurably advanced materials processing, purification of chemical elements, and the manufacture of pharmaceuticals and medicines. In addition, they have helped to preserve international peace by their ability to verify international arms control agreements.

In the future large space stations will serve as construction bases for specialized space objects designed to exploit space resources. The construction in space of a solar power system will make it possible to capture solar energy for transmission to earth. Other space-built satellites will allow for the mining of the mineral resources located on the moon and other celestial bodies. It has been suggested that such resources could be processed in the space environment so as to provide finished products for space use. Activities of this magnitude would call for the presence of a human population in space. It has even been suggested that the moon might become the situs of a permanent human habitation. Over time, as earth-based resources are diminished, while earth-based requirements are enlarged, a more pressing need for the exploitation and use of space-environment resources will arise.

Pending the proposal for operating space stations during the next decade, the operations being carried out and planned for the space shuttle and comparable foreign undertakings have necessitated the formation of an appropriate legal regime. Lessons learned will be directly applicable to space stations.

At present it is contemplated that the space station of the 1990s will be the product of cooperative efforts between the United States and friendly European countries. This prospect resulted recently in an international colloquium on "Space Stations, Legal Aspects of Scientific Commercial Use in a Framework of Transatlantic Cooperation" held in Hamburg, Germany, on October 3–4, 1984.[2] At its annual meeting of the International Institute of Space Law in Lausanne, Switzerland, October 8–13, 1984, several sessions were allocated to a consideration of the international and domestic issues and presented by the prospect of space stations.[3] The involvement of many States in joint space ventures of this magnitude will raise many political and legal issues. Some experience has been gained from prior cooperative undertakings between the United States and numerous foreign countries.[4] The infrastructure created by the European Space Agency will also provide many valuable insights.

Large-scale space station activities, which take place in a hazardous and inhospitable natural environment, will result in a number of physical problems, which can either be prevented or overcome through suitable internationally recognized practices and procedures. Concerns have been voiced respecting pollution, contamination, solid debris, the monitoring of debris, and collision probabilities. In October 1984, it was estimated that the United States was tracking up to 5,300 pieces of debris, almost all of relatively small size. However, there are now in orbit a number of inactive satellites, which pose danger for active and future space objects. These conditions have led to suggestions that through international agreement provision could be made for safe launch corridors and timely notices of launch, although at present the requirement of prior notice does not appear to be practical. If pollution is considered an excessive amount of space clutter and debris, then avoiding such a condition would be desirable. In a positive sense there is a common need to insure that space and its natural resources are used in the most efficient, economical, and equitable manner possible. Failure to make maximum use of space capabilities is as undesirable as physical harm to the natural environment. From this perspective a stable regulatory framework would prove the wisdom underlying the adage that an ounce of prevention is worth a pound of cure.

National Considerations

As the United States addresses its cooperative role in the establishment of large space stations it will be obliged to consider the municipal legal rules governing internationally sponsored launches and interpersonal relationships on the space station. It will also have to fashion domestic institutions charged with protecting the interests both of participating individuals and the larger public.

Among the substantive areas of law that will have to be dealt with are the internal public order of the space station, including the powers of the spacecraft commander, and the applicable criminal law, the protection of intellectual property, including copyrights and patents, the rights of non-nationals on the space objects, and wide-ranging jurisdictional problems, including the adoption of one or more legal principles, such as the nationality, territoriality, universality, protective, or passive personality principles.[5] One issue that may require specific attention is the U.S. position on monopolies and restraints of trade, which is addressed from a different perspective by European states.

In the United States at present there are a number of national departments and administrative agencies which possess separate mandates relating to space activity. At the departmental level are the Department of State, the Department of Commerce, and the Department of Transportation. The Department of Defense, including the armed services, also has a natural involvement in space station activities. In order to meet the increase in launching activities the Department of Transportation has created an Office of Commercial Space Transportation. This department also contains the Federal Aviation Administration. The Department of Defense now has a North American Aerospace Defense Command. In addition, the National Aeronautics and Space Administration and the Federal Communications Commission, among others, are critically involved in outer space activities. The development of a coherent set of legal rules relating to large-scale space activity will require coordination on the part of all these national instrumentalities. It may also be expected that state laws and authorities will have to change their laws and procedures to conform to federal mandates.

The need for the careful orchestration of a clear-cut domestic legal regime should be obvious. American firms are preparing to invest billions of dollars in the science and technology required to make space stations a reality. They are willing to accept the national challenge to explore, exploit, and use the space environment and its natural resources for peaceful purposes. They are being encouraged by present federal policies. To allow for the fulfillment of such expectations it is evident that the government itself, in addition to monetary assistance in suitable circumstances, should make a large investment in time and energy in order to perfect a suitable municipal legal regime. The economic rewards are likely to have a strong impact on the economies of all participants.

THE 1967 PRINCIPLES TREATY AND THE 1979 MOON AGREEMENT: THEIR APPLICATION TO THE COMMERCIALIZATION OF SPACE

A draft Agreement Governing the Activities of States on the Moon and Other Celestial Bodies went into effect in 1984, following its adoption by the UN General Assembly in 1979. Only five countries are now bound by the agreement: Austria, Chile, the Netherlands, the Philippines, and Uruguay. Both the United States and the Soviet Union played an active part in drafting the treaty. At the time it was presented to the General Assembly both countries approved it.[6]

The Moon Agreement is notable in several respects. It foresaw the need to restrict armaments on and around the moon, if commercial activities were to take place there and prosper. Thus, Article 3 provided that parties were not to "place in orbit or other trajectory to or around the Moon objects carrying nuclear weapons or any other kinds of weapons of mass destruction or place or use such weapons on or in the Moon,"[7] a reemphasis of the fundamental obligation contained in Article 4 of the 1967 Principles Treaty relating to orbits around the earth.

The 1979 agreement was equally notable in extending the mankind principle of Article 1 of the Principles Treaty, although with a different emphasis, and with a detailed recitation of goals, on the activities to be carried out in or on the moon. Article 11, paragraph 1, of the Moon Agreement stated that "the Moon and its natural resources are the common heritage of mankind."[8]

Prior to the compromise acceptance by the negotiators of the foregoing formulation many proposals were advanced regarding the legal status of the moon and other celestial bodies, particularly as they related to the appropriation and exploitation of the area and of the natural resources of the area. These views have been summarized:

> Some considered that such natural resources could be lawfully exploited; others viewed such activity as an unlawful appropriation. Among those who favored the legality of exploitation of resources were some who reserved this activity to States; others considered such activity to be lawful when pursued by both States and private legal persons. Some held the view that such exploitation should be restricted to scientific activity; others considered that the exploitation might be directed to both scientific and commercial needs.[9]

During the debates at the UN competing juridical doctrines were propounded. It was urged at one time that the United Nations should hold legal title to the area and its resources. One view advanced was that the area should constitute a *res nullius*, subject to the claim of exclusive sovereignty by a state or states. The contrary proposal was also put forward, namely, that the area should be a *res communis*, thereby open to the common uses of all potential explorers and exploiters, and not subject to a regime of exclusivity. The Argentinian space lawyer, A. A. Cocca, urged that the area be subject to a *res communis humanitatus* regime.[10] His proposal ripened into the outer space principle of the Common Heritage of Mankind (CHM).

During the search for a key principle governing the exploitation of the moon and other celestial bodies, including their natural resources, it was suggested that an analogy might be drawn between the regime for the continental shelf, which was founded on the exclusivity principle, and the regime for the moon and its natural resources. In the end the CHM principle was accepted, since it was based on the *res communis* principle. However, the CHM principle extended the *res communis* principle by calling for a new legal regime designed from moon resource exploitation. Article 11, paragraph 7 (d), in making provision for a new international regime provided:

> An equitable sharing by all States Parties in the benefits derived from those resources, whereby the interests and needs of the developing countries, as well as the efforts of those countries which have contributed either directly or indirectly to the exploration of the Moon, shall be given special consideration.[11]

This provision adopted the theme found in other international agreements in which equitable distributions are to take place. In order to clarify the meaning of "equitable" the United States has indicated in different international gatherings that this term, while conveying the views of fairness and justice, does not mean "equal."

Following the adoption at the UN of the Moon Agreement several American senators suggested that the CHM provisions could be damaging to the national economic and security interests of the United States.[12] This proposition was discounted by Secretary of State Vance, who correctly pointed out that the treaty, and in particular Article 11, paragraph 3, would "permit ownership to be exercised by States or private entities over those natural resources which have been removed from their 'place' on or below the surface of the Moon or other celestial bodies."[13]

To this it should be added that Article 11 stipulates that initially, i.e., in the early stages of the exploitative process, the parties are entitled to retain all of the benefits derived from the commercial uses of the moon and other celestial bodies. The geographical scope of this right extends to "orbits around or other trajectories to or around it."[14]

The agreement makes an important distinction between preliminary exploratory and use activities, during which time all economic benefits would flow to the explorer and user, and a more advanced stage of commercial activity. Thus, at such time as the original commercial activities were to ripen into true "exploitation" of natural resources, which has been taken to mean extensive or large-scale activities, the parties to the agreement are called upon to create a new international legal regime. It would be the function of such a regime to effect a distribution of the "benefits" derived from such large-scale exploitation in accordance with the formula set out in Article 11, paragraph 7 (d).

Such a regime is to be created, according to the terms of the treaty, only by those states that have ratified it. This means that neither the United States nor the Soviet Union would participate in the design of the new legal regime. Since the CHM principle would be applicable only to the participants in the new regime, as implemented by the newly established international organization, nonparties would continue to be governed by the modified *res communis* provisions appearing in Articles 1 and 2 of the 1967 Principles Treaty. Article 9, paragraph 2, of the Moon Agreement adopts the wide-ranging provisions of Article 1 of the Principles Treaty. Article 2 of that agreement provides that "outer space, including the Moon and other celestial bodies, is not subject to national apropriation by claim of sovereignty, by means of use or occupation, or by any other means."[15] Article 11, paragraph 2, of the Moon Agreement applies the same principle to the moon. In the light of these provisions it would be possible for the parties to the Moon Agreement to be governed by a CHM regime, while nonparties would still have to conform to the modified *res communis* regime.

In reaching a policy decision the United States as a space-resource State would have to consider whether it would have more to gain from the specific provisions of the Moon Agreement, including acceptance of the duty to share some of the benefits derived from exploitative activity, or from reliance on the *res communis* principle, which does not require, and has not resulted in, the sharing of specific benefits. While the Moon Agreement does not repeat the terms of Article 1, paragraph 2, of the Principles Treaty providing for the free and equal exploration, exploitation, and use of the space environment, this right is so fundamental that it would apply to moon activities in any case. Thus, in the future it would be possible, assuming that the United States does not become a party to the Moon Agreement, for it to have recourse to the *res communis* regime. At the same time the States that have ratified the Moon Agreement, none of which are now space-resource states, would be governed by the CHM regime.

This situation would be somewhat akin to the present United States position relating to the 1982 Law of the Sea Convention, to which the United States has not become a party. The 1982 agreement contains many customary principles of general international law upon which the United States can rely without formal acceptance and at the same time the United States has entered into side arrangements with other maritime countries on selected ocean matters.

A parallelism exists between the rights and duties of states relating to the exploration, exploitation, and use of the ocean and its resources and the space environment and its resources. The ocean beyond the limits of national sovereignty has been treated as a *res communis* area and tied to the expression "freedom of the high seas." The 1982 Law of the Sea Convention has established a CHM area, which is subject to constraints when states seek to dispose of benefits resulting from national exploitative activities.

The 1967 Principles Treaty relied by way of analogy on the *res communis* doctrine. This found expression in Articles 1 and 2 of the treaty. Article 1, in referring to the exploration and use of the space environment, provided that such activity was to be carried out for the benefit and interests of all countries and was to be the province of all mankind. Since this departed from the traditional view of *res communis*, the United States Senate Committee on Foreign Relations advanced the following understanding respecting the meaning of Article 1: "Nothing in Article 1, paragraph 1, of the treaty diminishes or alters the right of the United States to determine how it shares the benefits and results of its space activities."[16] The understanding was designed to reinforce the *res communis* principle. Nonetheless, Article 1, by its terms, was not in conformity with a strict view of the principle, since the article placed greater emphasis on the benefits and interests of all countries in the service of mankind than has the traditional *res communis* principle.

The 1979 Moon Agreement, while building on Article 1 of the Principles Treaty, went beyond it in adopting the CHM principle, even though, as noted above, this principle will not be applied until at some future date there is a large-scale exploitation of moon resources. The principle will not be implemented until after the parties to the agreement have established an appropriate institution having the authority to effect the distribution of benefits called for under Article 11, paragraph 7, of the treaty.

Thus, for both the 1982 Law of the Sea Convention and the 1979 Moon Agreement the *res communis* principle has been modified in the specific contexts of the two treaties, subject, as indicated, to future eventualities.

Moreover, just as the Law of the Sea Convention contained many instances of general principles of customary international law, so the 1967 Principles Treaty as augmented by the Moon Agreement also contained references to the traditional *res communis* principle, namely, the present right to explore, exploit, and use the space environment and its resources freely and equally, and to have free access to the area and its resources. Under such circumstances, particularly since the CHM provision constitutes an extension of the *res communis* principle, it would appear that States for policy purposes are free to give their support to the *res communis* principle as it applies respectively to the ocean and to the space environment. Their total commitment to the *res communis* principle would be evidenced by not accepting the 1982 and 1979 agreements. Further evidence would be the acceptance of an agreement or agreements in which their unrestricted support for the traditional *res communis* principle was made known.

While such national policies may be legally supportable, it should be noted that the CHM principle has received very wide-ranging approval. Since the resources that are or will be explored, exploited, and used are situated in the world's commons, it may be that the self-interest of the space-resource states would best be served through the acceptance of the principle. As in most matters the decision will depend on the precise manner in which the sharing of benefits formula of Article 11, paragraph

7, of the Moon Agreement is implemented. In my opinion an equitable sharing of the benefits would give the largest shares to the countries which have produced the benefits.

CONCLUSION

From the foregoing it may be concluded that states are being provided with a choice of international legal regimes for the exploration, use, and exploitation of the moon and other celestial bodies and the resources of these areas. Some eighty-five states, including all of the space-resource countries, have accepted the modified *res communis* principle contained in Article 1 of the 1967 Principles Treaty.

Of these eighty-five countries only five have become bound by the 1979 Moon Agreement, and these five are not space-resource countries. The entry into force of the Moon Agreement among these several states has not served to put the CHM principle into operation. This must await the practical and large-scale exploitation of moon resources. Thus, for the moment there is no practical conflict between the two international agreements. Nonetheless, it is important to realize that over time, and when and if several of the space-resource states become bound by the Moon Agreement, it will be necessary to accommodate the respective international legal regimes. Such an accommodation will be essential to the successful commercialization of space. Over time it is very probable that a new international intergovernmental organization will come into being to aid in maintaining a stable regulatory framework for the space environment and its natural resources.

NOTES

1. 18 UST 2410; TIAS 6347; 610 UNTS 205. The following illustrations are taken from the Treaty. It has been ratified by 85 States including the United States, the Soviet Union, and the Peoples Republic of China.

2. The papers presented at the Conference are to be published in "Studies in

Air and Space Law" under the auspices of the Institute of Air and Space Law, University of Cologne.

    3. The papers presented at the international scientific and legal round table and of the Institute are to be published in the Proceedings of the 27th Colloquium on the Law of Outer Space, American Institute of Aeronautics and Astronautics, New York.

    4. C. Q. Christol, "Space Joint Ventures: The United States and Developing Nations," *University of Akron Law Reviews* (1975), 8:398–415.

    5. C. Q. Christol, "The International Law of Outer Space" (Washington, D.C.: GPO, 1966), pp 416–418.

    6. Senate Committee on Commerce, Science, and Transportation, "Agreement Governing the Activities of States on the Moon and Other Celestial Bodies, "Parts 1–4, 96th Cong., 2d Sess.; "The Moon Treaty," Hearings before the Subcommittee on Science, Technology, and Space of the Committee on Commerce, Science, and Transportation, 96th Cong. 2d Sess (1980); C. Q. Christol, "The Common Heritage of Mankind Provision in the 1979 Agreement Governing the Activities of States on the Moon and Other Celestial Bodies," *The International Lawyer* (1980), 14:429.

    7. U.N. Doc. A/34/664, November 12, 1979; *International Legal Materials* (1979), 18:1434.

    8. *Ibid.*

    9. C. Q. Christol, "The Common Heritage of Mankind Provision," p. 447.

    10. A. A. Cocca, "The Principle of the 'Common Heritage of All Mankind' as Applied to Natural Resources from Outer Space and Celestial Bodies," *Proceedings of the 16th Colloquium on the Law of Outer Space* (1974), p. 174.

    11. U.N. Doc A/34/664, November 12, 1979; *International Legal Materials* (1979), 18:1434.

    12. U.S. Department of State, "Digest of United States Practice in International Law, 1979" (Washington, D.C.: GPO, 1983), p. 1172.

    13. *Ibid.* Secretary of State Vance observed that such removal is also permitted by Article 1, par. 2 of the 1967 Principles Treaty, which states, inter alia, that "Outer space, including the Moon and other celestial bodies, shall be free for exploration and use by all States."

    14. U.N. Doc A/34/664, November 12, 1979; *International Legal Materials* (1979), 18:1434.

    15. 18 UST 2410; TIAS 6347; 610 UNTS 205.

    16. Senate Committee on Foreign Relations, "Treaty on Outer Space," S. Exec. Rep. No. 8, 90th Cong., 1st Sess. (1967), p. 4.

T W O

Will Satellites and Optical Fiber Collide or Coexist?

ROY A. LAYTON

S atellite and optical fiber systems may be on a collision course. Announcements of planned satellite launches as well as announcements for millions of miles of fiber-optic circuits raise a serious question: Will sufficient demand exist to use such huge oversupply of transmission capacity? If all of the announced projects get off the ground, the total capacity of the long distance market will grow nearly eightfold over the next five years.

COLLIDING SUPPLY FORCES

Since price is a function of supply and demand, these colliding supply forces may create fierce competition and declining prices that could shake the foundation of deregulated telecommunications. On the other hand, innovation might create new demands extending beyond the range of applications in use today. Can we avoid a telecommunications San Andreas Fault?

The French poet Paul Valéry once remarked that the

future isn't what it used to be. His words stand as both a caution against history's capacity to blindside us and as a challenge to look ahead.

As satellite and fiber-optic supplies progress down similar paths, a look at the history of these technologies will serve as a benchmark for future growth comparisons. Comparing the facilities available today to the projected growth for 1990 will help you see the extent to which evolving technology will shape twentieth-century demand.

Looking at the historical perspective first, the paradox of the late twentieth-century telecommunications evolved from the genius of Alexander Graham Bell. Among his many patents, four covered his invention of the photophone and two the selenium cell. In 1880, Bell demonstrated that a beam of sunlight reflected off an acoustic horn-mounted diaphragm could be used to carry sound to a distant speaker that was equipped with a selenium photocell. In a single demonstration, Bell laid the foundations for lightwave transmission and remote solar-driven relays, which are components of satellite systems.

Art imitated research when, in 1945, science fiction writer A. C. Clarke first proposed satellite communications. In 1954, J. R. Pierce suggested that you could accomplish such communication by bouncing signals off passive satellites or relaying them through electronically equipped units. Pierce concluded that you could place a satellite in geostationary orbit so that its position would be fixed in relation to a ground point if its orbital velocity were to match the earth's rotation at 22,300 miles above the equator.

The Edge of Space

Three years later the Soviet Union astounded the world and partially realized Pierce's vision with the launch of Sputnik, a primitive orbital radio transmitter. The United States responded with an accelerated program that resulted in the successful 1960 Echo I experiment using a 100-foot diameter passive, low altitude reflection balloon.

Following construction of the advanced earth station

at Andover, Maine, in 1962, the Bell System launched its first Telstar satellite, ushering in the age of the full communications relay satellite. Further milestones in the development of satellite technology for commercial communications include:

- First satellite in a series of launches by the international consortium INTELSAT, organized in 1964 to develop a world satellite network. This series of satellites refined the technology later used to produce Westar I and most satellites in use today. (1966)
- First geosynchronous orbit commercial communications satellite launched by the Canadian Government—ANIK-A. (1972)
- Western Union launched what Newsweek president Gibson McCabe called "The Golden Spike of the American Skies." The launch provided an alternative to AT&T for long-distance voice, data, and video telecommunications and served to rejuvenate the cable television industry. It provided a new source of programming for broadcast stations, revolutionized electronic news gathering, and set the stage for satellite broadcast to homes—Westar I. (1974).
- Satellite Business Systems (SBS), a venture of IBM, COMSAT, and the Aetna Insurance Company launched the first commercial communications satellite primarily designed for use by private earth stations—SBS-I. (1980)
- The Spaceshuttle Columbia pioneered the concept of a reusable "space trunk" for satellite launch, maintenance and replacement with its first communications satellites—SBS-III and ANIK-B. (1984)

A $3.1 billion satellite industry has emerged from the U.S. space program. The market-share players: the American Satellite Company, a partnership between Continental Telecom, Inc., and Fairchild Industries; RCA; SBS; Western Union; Telstar (AT&T); Galaxy (Hughes); and Comstar (COMSAT).

Harnessing Light

Just as the pace of satellite service development was quickening with the formation of INTELSAT during the mid-1960s, a breakthrough occurred in harnessing light for future telecommunications applications. At ITT's Standard Telecommunications Laboratory in England, C. K. Kao and G. A. Hockham suggested that light waves could be guided by glass to where they were needed, solving the distance limitation of Bell's photophone. By 1970, scientists at Corning Glass Works had made the idea work.

Hair-thin pieces of silica glass could bend easily to serve as "waveguides" for light waves. At the same time, advances in semiconductor technology made possible the fabrication of efficient light sources which could be modulated with an external digital signal. The convergence of these two technologies through field trials starting in 1975 harnessed light as a source for telecommunications transmission.

In an optical fiber, light is funneled in one end, is repeatedly reflected at a low critical angle off the walls of the fiber, and emerges at the same angle at the other end—as if it had been placed in a pipeline. In contrast with our everyday experience with sunlight, this property of optical fiber holds true no matter how many turns and twists the fiber makes along its length.

A fraction of the diameter, lighter weight, and circuit-for-circuit cheaper than copper wiring, fiber-optic cable provides transmission capacities beyond any conventional medium. A single pair of glass strands can transmit more than 14,000 32k-bps voice conversations, more than 50,000 data channels, or a variety of voice, data, and video combinations at 560M bps, limited only by user need and matched electronics at each termination point.

For nearly two decades the promise of transmitting information over fiber hovered on the telecommunications horizon. By 1978 some 600 miles of fiber optics had been installed worldwide.

Confronted with the realities of the marketplace and recognizing the potential of fiber optics, many companies concentrated on the growing telephone market. Significant milestones in this development include:

- General Telephone installs nearly six miles of fiber for telephone service between Artesia and Long Beach. (1977)
- Illinois Bell connects two central offices approximately one mile apart. (1978)
- AT&T announces a 611-mile system that could connect major cities in the Boston-Washington corridor. (1980)
- MCI announces a New York to Washington link along Amtrak right of way. (1983)
- Lightnet, a venture between CSX and Southern New England Telephone, announces initial plans to lay fiber alongside 5000 miles of railroad track. (1983)
- AT&T announces a $2 billion construction program to expand its digital communications network. Included are 21,000 miles of fiber-optic cable. (1984)

The possible collision between satellite and fiber systems was interrupted when a five-year-old antitrust suit between AT&T and the Justice Department was settled on January 8, 1982.

With a target date of January 1, 1984, AT&T and the Department of Justice settled what had been the largest antitrust suit in history and agreed to sweeping changes in the telephone marketplace. Life for America's telephone users changed irrevocably.

The Modified Final Judgment (MFJ) ruling opened the door fully for competition and, on both the supply and demand sides, its provisions removed the major economic structural barriers that had existed in the industry. The cornerstone of the settlement—equal access to local facilities—established competition as the substitute for regulation of interstate facilities. Clearly, the forward trend in U.S. telecommunications is competitive, cost-effective transmission at prices forged in the crucible of supply and demand.

Aftershocks from divestiture further destabilized a telecommunications industry still recovering from what Prime Minister Thatcher's former aide, Sir John Hoskyns, termed "insurmountable opportunities."

The end-user market has been placed in a four-way crossfire between AT&T's public network, established microwave, new competing satellite, and a growing number of fiber-optic carriers. Claims and counterclaims abound over the application superiority of each respective technology. And prospects of economic gain to "buy in" attract users of long-haul services.

COMPARATIVE ADVANTAGES OF SATELLITE AND FIBER OPTIC TECHNOLOGIES

Satellites and fiber optics each have salient technological and economic advantages which will bear on future competition for international transmission.

Satellites are insensitive to distance. All transmissions travel 44,600 miles via spacelink and are insensitive to distance between uplink and downlink locations within the footprint. Operating locations can be changed as business requires. This allows freedom from concern over available terrestrial facilities and long lead times.

Additionally, with only a few switching points between locations, compared to the hundreds of relay points involved in using terrestrial facilities, satellite signals stay neat and clean. There is much greater signal clarity than for conventional transmissions which tend to pick up noise.

Fiber optics, in comparison, allow freedom from the propagation delay which has been problematic for satellites. Fiber transmissions travel point to point at a fraction of the time it takes for the satellite transmission's round-trip distance. No pause between voice intervals occurs, nor are there lapses in data transmission.

Fiber optic transmission is not weakened by heavy rain and snow problems that affect satellite earth stations and nor is it affected by sunspots. Fiber optics have interference immunity, avoiding the radio, microwave, and airport interference that can cause site selection or hardware location problems that exist for C-band earth stations. Moreover, optical fiber eliminates costly foundation or roof reinforcement. All commercial buildings have three or more rights of way—telco, power, fuel—that can be used to install the fiber cable.

Cost considerations will no doubt continue to be relevant to decisions about these two technologies. The potential cost savings for satellites are greatest when the distance between serving points exceeds 1,100 miles compared to microwave, 120 miles when compared to fiber for DS-1 equivalent increments. The potential cost savings for fiber optics, on the other hand, are greatest when the distance between serving points is less than 700 miles compared to microwave, and 2,500 miles compared to satellite for carrier increments of 2,700 circuits.

A Surplus of Satellites

A surplus of satellite transmission capacity continues to define the marketplace. According to the FCC Common Carrier Bureau, twenty-one domestic satellites are now operating and ten more are scheduled to be launched this year. It has been estimated that only two-thirds of current domestic capacity is being used.

Transponders have moved from the attempted auction prices of 1980 to the $3.5 million to $4 million each that is rumored to be the prevailing rate. Compared to a transponder price of $13 million in 1981, the current rate means an opportunity era for entrepreneurs with new applications such as Equatorial Communications and end users who welcome declining prices.

Firms comprising the industry's brain trust have generated a stream of research reports projecting supply, demand, and comparative costs. Among these, a projection by International Resource Development shows the change in use of domestic satellites from 1983 to 1990. Of 356 transponders in 1983, 53 percent were used for voice, 28 percent for video, and 19 percent for data. By 1990, voice is projected to be 43 percent of 1,050 estimated transponders, data will increase to 40 percent, and video will drop to 17 percent.

The basis for evaluating projected capacity is the total

circuit miles available at year-end 1984. According to Compucon Inc., AT&T totaled 811 million circuit miles, MCI's microwave, cable, and fiber totaled 265 million circuit miles and GTE's network comprised 100 million circuit miles, for an overall total of 1.18 billion circuit miles.

In addition, NASA reported 356 operational transponders, of which 72 percent or 256 were in use for voice and data transmission, according to International Data Resources. If you use a ratio of 300 circuits per transponder and an average of 1,000 miles per "equivalent" transponder circuit, you add an additional 76.8 million circuit miles of satellite capacity to the terrestrial capacity just outlined. Therefore, the total terrestrial and satellite capacity at year-end 1984 is estimated at 1.25 billion circuit miles.

Several estimates of transponder growth are available. International Resource Development projects 1,050 by 1990. James Martin, speaking at the Communications Network 1985 conference, projected 1,072 in 1987. NASA projects 1,145 by 1990. These figures are well within the 1,254 transponder capacity available at two degree spacing within the orbital arc. If you estimate transponder availability at 1,000 in 1990 and convert it to "equivalent" terrestrial capacity, the result is approximately 300 million circuit miles of satellite capacity available in 1990.

Fiber capacity will grow sixfold by 1988 if approximately $6 billion worth of announced projects are installed. For example, United Telecommunications plans a 23,000-mile system. MCI and GTE have announced plans that total 22,000 miles. AT&T announced a 21,000-mile network. Fibertrack, a joint venture of Norfolk Southern Corp. and Santa Fe Southern Pacific Corp., plans an 8,100-mile network. Lightnet plans a 4,000-mile system serving 43 cities. Many other projects have been announced and more are on the drawing boards.

Transmission Capacity in 1990

By 1990, it is conceivable that 8.19 billion circuit miles of terrestrial capacity could collide head-on in the marketplace with 300 million circuit miles of satellite capacity. If total capacity approaches 8 billion miles by 1990 as estimated, then capacity will shift from S to S1. Demand is expected to increase at a lower rate than supply and thus would shift from D to D1. As a result, the market price per circuit would gradually sink along an arc, Point A to Point B, determined by the relative time for the supply and demand shifts to occur.

Synthesizing the supply and demand projections, prices should decline gradually into the early 1990s as competition intensifies and new applications fail to keep up with supply. Ultimately, market forces will take care of the imbalance.

A Sampling of Opinions

"You have to look at this in two time frames," contends Ray Fentriss, executive vice president of Gartner Group.

For this decade, satellite transmission should continue to be viable and competetive. Beyond 1990, the emerging fiber-optic facilities will provide ubiquitous high bandwidth capabilities for the highest volume traffic routes. Even if fiber proves to be more economical on the heavy traffic routes, satellites will enjoy success in providing transportable earth stations that allow instant connectivity and broadband capacity where no other communications facilities exist.

These two technologies can co-exist in the future and their unique characteristics will be exploited to meet the growing need for pervasive, easy-to-connect bandwidth.

At the Newport Conference on Fiber-optic Markets, Charles Wakeman, vice president and general manager of Siecor Corp., predicts a hesitation in the market, a downturn in the growth curve between 1987 and 1990 as dust settles from the competitive battles now starting.

James Martin, widely regarded as the foremost spokesman on the social and commercial impact of computers in communications, said at the Communications Networks 1985 conference, "I doubt whether we will have a serious bandwidth glut in fiber or satellite. When technology permits taking the brakes off 9.6 transmission, the need for faster speeds will require large amounts of bandwidth."

"With the advent of fiber-optic technology, many people believe that it has become a ubiquitous communications link for long-distance traffic and that the role of satellite technology is declining," says Arthur Parsons, senior vice president of American Satellite. "In actuality we've seen very little fiber laid and only along high-density routes which can support its high cost."

INFORMATION AGE OPPORTUNITY

"In the future," Parsons says, "fiber will complement satellite technology and a trend toward integrated networks which use fiber for short distances will occur. Satellites are uniquely suited for such services as broadcasting and it will be virtually impossible to replace them. As larger satellites with enormous transmission capacity and power are launched, space segment cost will drop dramatically and antenna sizes will shrink. I predict a very bright future for satellites in the business communications marketplace."

Howard Anderson, managing director of the Yankee Group, says, "The only way we will be able to use all of this capacity is if Congress passes a law requiring every American to participate in five videoconferences a day."

As you evaluate these opinions, remember the words of John W. Gardner: "We are all faced with a series of great opportunities brilliantly disguised as insoluble problems."

The United States already leads the world in research and development, spending more than $70 billion per year. Communications and electronics are among the top five investment areas.

As the domestic economy moves forward from the industrial to the information age, telecommunications is fast becoming the new strategic variable for every business enterprise. Entrepreneurial spirit will raise capital and build capacity. Visionaries will develop new applications not conceivable today. The fault plane in the foundation of deregulated telecommunications should remain stable as the forces of technology and growth propel demand to fill new satellite and fiber systems. The race is on.

Reprinted with permission from *Telecommunication Products Plus Technology*, (Littleton, Mass.: PenWell).

PART II:
THE EMERGENCE OF SPECIALIZED AND REGIONAL SATELLITE SYSTEMS

T H R E E

Competition by Private Carriers in International Commercial Satellite Traffic: Conceptual and Historical Background

MARCELLUS S. SNOW

The aim of this essay is to provide a historical and theoretical context from which to judge the present controversy regarding competition in North Atlantic commercial satellite services. The approach will be to supply the reader with both information and methodologies on which to base a decision about this important policy issue, rather than to make such a decision on the reader's behalf.

HISTORICAL PERSPECTIVE

This section explores important policy actions regarding competition and private entrepreneurship in the provision of telecommunications facilities and services, many of them directly involving international satellite traffic.

The Communications Satellite Act of 1962

The first artificial satellite was launched by the Soviet Union in 1957, and the United States soon followed suit. As satellites became commercially viable for communications purposes in the early 1960s, Congress considered whether private or public ownership and operation would be preferable. There was no debate about "monopoly" per se—the scarcity of the new resource made the selection of a single provider a foregone conclusion.[1] Instead, debate centered around the question of whether the monopoly should be awarded to private interests or retained by the Federal government. Arguments for Federal ownership stressed the unknown aspects of the infant technology and warned against a "sellout" to private interests.[2] Those favoring private ownership, in turn, disagreed with one another as to whether the satellite enterprise should belong to existing overseas carriers—AT&T, Western Union International, RCA Globcom, and ITT—or whether the public at large should own the shares. After acrimonious debate, private ownership was decided upon, and as a compromise half of the stock of the newly formed Communications Satellite Corporation (Comsat) was purchased by existing carriers, while the other half was sold to the public at large. The carriers sold their interest during the 1970s.

Comsat, then, was awarded a monopoly in U.S. commercial satellite traffic overseas. The technology was too new to consider any domestic applications, or the possibility of two or more firms competing for the overseas market. With demand and supply at very low levels, economies of scale seemed strong enough to argue for entrusting the market to a single carrier.

The INTELSAT Agreements

The original INTELSAT agreements concluded in 1964 were negotiated at American instigation among nineteen mostly industrialized countries, basically those with earth stations or the prospect of soon obtaining one.[3] The dramatic successes of the early transatlantic television transmissions made possible first by experimental and then by INTELSAT satellites increased the number of countries wishing to accede to the agreements. Thus, the original agreements were made provisional in nature to accommodate those countries wanting to share in the technology but not desiring to perpetuate the predominant position of the United States in satellite, launch, and earth station technology. Overall policy direction as well as medium-term management was therefore temporarily vested in an Interim Communications Satellite Committee (ICSC), on which voting quotas reflected investment shares. Day-to-day operation was contracted out to Comsat, acting as INTELSAT's manager.

The permanent or definitive agreements were arduously renegotiated during the years from 1969 to 1971 and entered into force early in 1973.[4] These replaced the ICSC with a Board of Governors which still votes by investment shares, but on which smaller users have more rights of representation and greater protection against undue influence by a small number of large users. In addition, two overarching membership structures were created for long-range policy matters, each having one vote per member: the Assembly of Parties, composed of the states that had signed the intergovernmental agreement; and the Meeting of Signatories, consisting of operating entities designated by those governments.

Two elements of the INTELSAT agreements are important for present purposes. First, INTELSAT was established along cooperative lines, in both the technical and, one can argue, the informal sense of the word. As an economic cooperative of investors and users, INTELSAT was financially structured so as to balance the investors' desire for high tariffs and the users' incentive for low tariffs.[5] By periodically aligning investment quotes with recent past usage shares for each signatory, INTELSAT makes the tariff level technically irrelevant. For cases of temporary imbalance of ownership and usage, as well as for non-using investors and nonmember users, the tariffs are set at a level that pays INTELSAT members a cumulative annual return of 14 percent on their net investment. Reflecting dramatic advances in technology and increases in usage, the original annual rate of $32,000 for a single voice-grade channel has now dropped to below $5,000.

A second issue in renegotiating the agreements, crucial to the purpose of this article, involved the prohibition of systems competing with INTELSAT, Article 14(d). By the late 1960s, such a threat had already become commercially feasible. The Canadian domestic system was already in operation, Indonesia's Palapa system was being designed, and the regional systems in Europe and among the Arab states were under consideration. In the original agreements INTELSAT's members had awarded the organization exclusive rights to operate "the" global commercial satellite system, although in the context of the mid-1960s the possibility of credible competition was still remote. This had changed five years later, and the issue of separate systems was easily the most controversial during the process of renegotiation.

The Authorized User Decision and Its Reversal
The Communications Satellite Act referred to "authorized users" and "authorized entities." Since these were further unspecified in the act, many large user groups sought to obtain access to INTELSAT directly through Comsat rather than first going through the international common carriers. The FCC determined in its 1966 Authorized User Decision not to allow this bypass, declaring Comsat's role that of a "carrier's carrier."[6] At issue was whether Comsat would be allowed to introduce more competition into the U.S. market for overseas carriage by INTELSAT. The international carriers argued that they were at a disadvantage in any such competition, as they could not directly approach INTELSAT but had to go through Comsat. The decision reflected the extreme market segmentation orientation of the FCC in the mid-1960s, generally under the influence of the large carriers whose representatives still sat on Comsat's board of directors. Recently, in the interest of competition and deregulation, the FCC has overturned the Authorized User Decision, and Comsat can now compete directly with certain large users for INTELSAT traffic.[7]

The President's Task Force on Communications Policy
Late in the Johnson administration, the President's Task Force on Communications Policy, chaired by Eugene Rostow, issued its final report. It urged continued support of INTELSAT as a single global system. In addition, requests had been filed with the FCC since the mid-1960s to establish domestic satellite systems in the United States, and the Task Force addressed this issue. Again

on the grounds of economies of scale in production and management, it recommended that Comsat be selected as the sole purveyor of domestic satellite services.[8]

While the Task Force's report became an instant dead letter after the inauguration of the Nixon administration a month later, it is significant in being one of the first policy documents to make explicit use of the "natural monopoly" argument, at that time equivalent to the existence of scale economies, to justify prohibiting competition in satellite facilities and services.

"Open Skies" in U.S. Domestic Satellite Services

Early in 1970 the Nixon White House issued a memorandum to the FCC chairman urging exactly the opposite policy, on the basis that "no natural monopoly conditions appear to exist in the provision of specialized communications via satellite." Further, it proposed to allow "competition (in providing a domestic satellite system) to act within well-defined limits necessary to preclude anti-competitive practices and to assure that the competition works toward the public interest."[9]

As later implemented by the FCC, this policy became known as the "Open Skies" approach to domestic satellite systems in the United States. In opposition to the chosen entity approach used with INTELSAT internationally, the FCC implemented a policy of almost maximum competition in the domestic market. As a result, a number of systems have evolved gradually since the early 1970s, and by all accounts have provided a broad range of new and conventional services at acceptably low prices. This differs from the domestic policy of a number of countries, such as Canada and Indonesia, where early government entrepreneurship rather than market forces determined the pace of introduction, usually through a single entity. Other countries, however, notably Japan, Australia, and the United Kingdom, have adopted the evolutionary, market-based approach of the United States, albeit with less competition.

INTELSAT's Domestic Transponder Leasing Program

Article 8 of the INTELSAT Operating Agreement requires, or has been interpreted to require, that average-cost pricing be the guiding rule of the global system's tariff structure. This policy was confirmed by the ICSC's Finance Subcommittee as early as 1971, before the Operating Agreement had entered into force.

INTELSAT's policy decision to begin leasing whole, half, and quarter transponders for domestic usage on a preemptible basis—a policy still essentially intact today—was adopted in the early 1970s, when the organization faced a situation of considerable excess capacity under the average-cost pricing regime. Today, INTELSAT leases transponders to thirty-one different countries for a broad range of domestic purposes, including drilling platforms at sea, communications with noncontiguous national territory, and general economic development programs.

One can interpret the purpose and motivations of INTELSAT's transponder leasing policy from a number of viewpoints. It can, for example, be seen as an exercise in marginal-cost pricing. This, incidentally, would not be consistent with the profit-maximizing behavior usually attributed to a monopolist, which is to produce at the level for which marginal cost equals marginal revenue. More plausibly, perhaps, it can be regarded as a type of value-of-service pricing (price discrimination). This means that the value of a service to the user (as measured by the price elasticity of demand of its consumers) is inversely proportional to the price charged—the assumption being that domestic customers are more responsive to price than international users.

Aside from these more technical interpretations, INTELSAT's transponder leasing decision can be seen in a broader policy context. First, it can be regarded as an effort to meet the needs of the developing and European countries, which constitute the bulk of transponder lessees; each group, for often conflicting reasons, had been critical of vestiges of United States domination in INTELSAT in the early 1970s. For purposes of this essay, the most relevant interpretation of the transponder leasing decision would be to consider INTELSAT's action as an attempt to forestall the establishment of separate domestic or regional systems that would otherwise have accommodated the traffic that INTELSAT was subsequently able to attract. Thus the threat (as opposed to the reality) of competing separate systems was able, in this view, to galvanize INTELSAT to depart quite radically from the average-cost pricing principles enshrined in its Operating Agreement. The ability of potential competitors to enter and exit a market at relatively low cost and thus to affect the behavior of the monopolist or incumbent firm—even if actual entry does not occur—is a salient feature of the theory of contestability of market structure to be examined in part 4, where it will be applied to the North Atlantic competition issue.

Recent FCC and Congressional Attitudes

In this final section of part 2 we survey two recent committee reports which convey the mood of Congress and summarize the position of the FCC toward competition in international telecommunications. Issued during the early part of the first Reagan administration, they are indicative of the strong support that deregulation of the telecommunications industry has found in the United States. While the work of both committees was completed before the specific issue of competition for INTELSAT's North Atlantic routes arose, the reports serve to illustrate the technological, economic, political, and often ideological environment within which American policy toward such competition will be developed, chosen, and carried out.

Late in 1981, the majority staff of the Subcommittee on Telecommunications, Consumer Protection, and Finance of the House Committee on Energy and Commerce issued "Telecommunications in Transition: The Status of Competition in the Telecommunications Industry." Among its major conclusions:

> There is general agreement that a more competitive environment . . . is desirable . . . Recently, because of the major changes in technology, there is the possibility that today's limited alternatives can develop to the point where competitive market forces will govern the industry.[10]

The Senate Committee on Commerce, Science, and Transportation issued a report in the spring of 1983 entitled "Long-Range Goals in International Telecommunications and Information: An Outline for United States Policy."[11] While the House report responded primarily to the domestic situation, including attempts to rewrite the Communications Act of 1934, the Senate paper was set in a more international context and seems more defensive and ideological in nature. The report was prepared by the staff of the National Telecommunications and Information Administration (NTIA) of the Department of Commerce.

The NTIA report asserts in passing that INTELSAT is a "triumph" of U.S. foreign policy and then discusses the issue of competing "regional" satellite systems—a designation that indicates the wholly unanticipated nature of proposals for private North Atlantic satellites financed by U.S. entrepreneurs that were to reach the FCC soon thereafter. Competition today, it notes, is "a reality in the U.S. domestic, if not international, satellite market."[12]

The report does not address the apparent conflict between INTELSAT as a U.S. foreign policy "triumph" and the possible symbolic and economic damage that separate traffic and systems—supported by current American deregulatory policy— might inflict on that organization. This is a theme that is constantly used by INTELSAT officials in opposing separate North Atlantic facilities.[13]

As these policy documents suggest, there is ample evidence that in examining the North Atlantic traffic controversy, the United States will be animated more strongly by technical arguments and political beliefs regarding the efficacy of competition and free markets than by the real and symbolic achievements that INTELSAT represents for U.S. foreign policy.

ARGUMENTS AGAINST COMPETITION IN THE NORTH ATLANTIC COMMERCIAL SATELLITE MARKET

This section examines the economic and foreign policy considerations that have been advanced for maintaining monopoly against competitive entry. Some of these arguments are quite general and abstract, while others have been adapted specifically to the issue at hand. Most, however, will be discussed in the framework of telecommunications markets.

Natural Monopoly, Economies of Scale and Scope,
and Sustainability of Monopoly

By far the most common defense of monopoly in any public utilities market is the alleged presence of what is usually called "natural monopoly." Traditionally, this has often meant nothing more or less than economies of scale in a single output: the average cost of production declines, over the relevant range, as output increases.

A more recent concept used to define natural monopoly is that of cost subadditivity, emanating from the sustainability of monopoly/contestability of market structure literature.[14] Costs are said to be subadditive when the cost of producing a total amount of a single output is less than the cost of having two firms produce that amount together.

In a single-output setting, economies of scale and cost subadditivity are the same. Rarely, however, does a firm produce only a single output, particularly in the telecommunications sector. Most large carriers, for example, provide both video and data services along with telephony. Even telephony cannot be considered a single commodity, since it can be classified into submarkets based on time of day, length of transmission, route density, and so forth. Although there is a temptation from an engineering perspective to assume that telecommunications providers supply a single output called "bandwidth" or perhaps "bits of information," a more fruitful approach is to differentiate markets and products whenever variations in the price elasticity of demand are observed, as between, for example, residential and commercial users.

For decades, policy analysis of "natural monopoly" in public utilities was chained to the unlikely assumption of a single output. Since the mid-1970s, analysis of the multi-product case has resulted in a number of striking new insights. An important one for this discussion is that of economies of scope. Natural monopoly is seen as occurring under two quite separate conditions. Either two smaller firms produce the same mix of products but on a smaller scale, or two smaller firms completely specialize in one of the two outputs. In this second case, we say that the cost function exhibits economies of scope, meaning that a single producer can produce more cheaply than two firms, each specializing in one of the outputs.

Economies of scope, akin to the older concepts of joint and common costs, reflect complementarities in the production process. They are vital in considering the nature of the cost function and its implications for regulatory policy in the face of claims regarding the presence of "natural monopoly." Before turning to the case of INTELSAT and competing systems, we must add one final element to our conceptual tool kit, that of sustainability of natural monopoly.

In a single-output case, a firm with natural monopoly—economies of scale—need not worry about market entry by competitors. Any firm producing at a lower scale will incur higher average prices and can be undersold by the incumbent firm, thus eliminating financial incentives for small scale entry. Thus, the traditional rationale for government regulation of single product monopolists has been to protect the public from profit-maximizing behavior on the part of the monopolist rather than to protect the monopolist from entry.

In the multi-product case, however, an incumbent firm enjoying cost subadditivity or natural monopoly—and thus generally benefiting from economies of both scale and scope—might still fail to prevent profitable entry by rivals. Such competitors would typically choose to produce a proper subset of what is offered by the incumbent, for example, by specializing in one or more individual product lines. In other words, although a single firm—by virtue of cost subadditivity—is always able to offer a given market basket of outputs at least cost, there exist cases in which rival firms still have financial incentives to enter the market and compete for part of the incumbent's business. When this happens, the incumbent must cut back production, and the overall cost of the same total market output rises. This is the case of an unsustainable natural monopoly. If no such incentives for profitable entry exists, the monopoly is said to be sustainable.[15] The existence of unsustainable multi-product natural monopolies is a vital public policy question for regulatory authorities and has important implications for INTELSAT.[16]

INTELSAT is clearly a multi-output enterprise. Its service offerings are differentiated by technical features (voice, data, video), by route, by region (Atlantic, Pacific, and Indian Ocean), by user restrictions (preemptible transponder leases), and by many other criteria. Does it have economies of scale and scope? Is it a natural monopoly, and, if so, is it a sustainable one?

One important point often cited in defense of INTELSAT is that it represents only about 10 per cent of total satellite communications costs, the remaining 90 per cent residing in earth station and various terrestrial transmission expenses.[17] Reference is often made to the claim that from an engineering point of view, system "optimization" would be possible only if the same entity operated both the earth and space segments. What is probably meant by this is that there are economies of scope in providing both earth and space segments through a single entity rather than through over one hundred (INTELSAT and 108 signatories, each of the latter with its own earth segment). Because of the political and institutional impossibility of unifying the earth and space segments—which is not at issue in the North Atlantic route debate—we will neglect this rather obvious source of potential economies in what follows.

Even in the space segment, it is conceivable and perhaps probable that INTELSAT enjoys a multi-output natural monopoly and economies of scope.[18] Suppose that there are three homogeneous outputs which we will call voice, video, and data service, and that INTELSAT presently supplies respective levels x, y, and z of those services. Cost subadditivity, the salient property of its natural monopoly, would then assure that no combination of two or more firms—presumably including INTELSAT—could provide output bundle (x,y,z) at lower cost. Assume, however, that INTELSAT's natural monopoly is unsustainable. One consequence of that unsustainability could be that two competing firms or systems, specializing in video and data, respectively, might find financial incentives to compete for all or part of INTELSAT's business in those services. In the case of complete specialization, we would then have the output vector $(x,0,0)$ for INTELSAT, $(0,y,0)$ for the video firm, and $(0,0,z)$ for the data enterprise. Each would be earning a profit and would thus have a financial incentive to remain in the market. Yet cost subadditivity would assure that the same output after entry is produced at a higher overall cost, to the presumed detriment of users as a whole. The important public policy issue is whether, under such circumstances, INTELSAT's unsustainable natural monopoly should be artificially sustained by entry restrictions and/or moral suasion by INTELSAT's membership, based on positive findings of "significant economic harm" under Article 14(d). To be sure, we have established a case for economic harm to satellite users as a whole. A case for harm to INTELSAT itself would have to be based on the consequences of losing certain categories of traffic to higher cost competitors as well as on a lack of diversification of output. This might be difficult to do in an environment of exponentially rising traffic.

There are other possibilities as well. Even though technical properties of the geostationary orbit or the earth's terrain may afford economies of scope in multi-region or global satellite services, the natural monopoly on which such economies are based might be unsustainable.[19] Subglobal systems, perhaps specializing in particular ocean regions or other geographic areas, would then have financial incentives for market entry, even though subadditivity would assure that any given combination of regional outputs would be more expensively provided by two or more systems than by the global system alone.

Finally, in a cooperative of owners and users, such as INTELSAT, the threat of competition can come from within as well as from outside. Individual user groups within an unsustainable natural monopoly can secede and create their own facilities or systems more cheaply for themselves but to the detriment of users as a whole. This may well have been the cost dynamic behind the establishment several years ago of INMARSAT, the global maritime satellite system. Perhaps heavy users of maritime communications, consisting of a small subset of INTELSAT members plus the Soviet Union, perceived that it was possible to specialize in the maritime satellite market profitably, although INTELSAT may have been able to provide any given vector of both maritime and public services at a lower price. Similar considerations might explain INTELSAT's disinclination or inability to specialize in other areas, such as aeronautical or land-based mobile services.

It should be clear that any of the questions posed regarding the existence and extent of natural monopoly, economies of scale and scope, or sustainability of natural monopoly for INTELSAT must depend for their answers on the existence and availability of complete, reliable, and accurate cost data and on the correct specification and estimation of the relevant cost functions. This is a task of urgent priority if important policy issues are to be resolved on a reasonably objective basis. Economies of scale, for example, have been estimated for the first decade of INTELSAT operation; in addition, more recent studies are extant.[20]

Competition in Services Rather Than Facilities

INTELSAT may find it in its interest to make facilities available for lease to firms that resell them, providing what are called value-added or resale services in a slightly different domestic context. This would be an alternative to establishing separate transmission facilities to furnish such services. In this way INTELSAT would retain its facilities monopoly but would move down the marketing chain to the status of a wholesaler in some of the services for which its facilities were ultimately used.

While the economic issues are difficult to sort out here, one might argue as follows. By providing circuits to wholesalers for later resale as value-added services, INTELSAT could retain the benefits of economies of scale. If separate facilities were established, by contrast, this would cause a loss of economies of scope. If INTELSAT has a natural monopoly that is strongly unsustainable, it might be possible for competitors to invest heavily in duplicate or parallel facilities and still have a financial incentive for entry. Perhaps an analogue of the indefeasible rights of usage provided to cable users could be established for certain categories of INTELSAT customers wishing to resell their circuits over an extended period of time.[21]

If separate facilities are uniformly opposed by INTELSAT members, the concept of value-added circuits is unacceptable to most telecommunications providers outside the United States.[22] This might pose a political barrier to their accommodation on

INTELSAT facilities and a greater spur to completely separate systems, again to the detriment of the user community as a whole.

Habit, Preference, and Stability

Recent studies of international telecommunications deregulation have included conjectures about preferences for the status quo based upon plausible organizational and psychological motives. In Australia, for example, a substantial measure of price stability and predictability may well be preferred by most of the population to prices that fall erratically in an environment of deregulation.[23]

Distributional matters are also paramount in deregulatory questions. Even when economic welfare as a whole increases, the welfare of certain individual user groups (the poor, rural customers, low-volume users) may well decline. Explicit subsidies to correct these difficulties are often proposed by economists. Direct assistance is more efficient in a purely technical sense than is the retention of the subsidy pattern implicit in most telecommunications pricing schemes, but is often politically impracticable.

Part of INTELSAT's unanimous opposition to competing North Atlantic facilities may well come from the fact that most of its signatories are either PTT administrations or other entities in telecommunications ministries, with various organizational and psychological motives for opposing change. These motives may very well have validity in terms of human and material resources, and should be considered to the extent possible in the overall calculus of costs and benefits brought to bear on the ultimate policy decision.

Cooperation Instead of Competition

Among the many non-Marxian socialist approaches to economic organization, the cooperative movement still retains some of its nineteenth-century appeal and following. INTELSAT, as noted, is technically an economic cooperative of owners and users.

Spokesmen for INTELSAT have argued against the introduction of competition, citing the harmony and cooperation that INTELSAT has displayed in its twenty years of efficient and apolitical existence.[24] Indeed, there is a sentiment toward INTELSAT among its members and many American proponents much akin to the supportive attitude toward AT&T before the first big competitive incursions by Microwave Communications, Inc. (MCI) in the 1960s or before its recent divestiture.[25] Part of this attitude can be summarized by the aphorism, "If it works, don't fix it."

Nonetheless, however strong the verbal appeal of the argument for cooperation instead of competition, it does not have anywhere near the theoretical buttressing of the natural monopoly arguments set forth above.

The Cream-Skimming Argument

Another defense of INTELSAT against competitive incursions, one employed with particular frequency on the high-traffic North Atlantic route, is that of cream-skimming. It goes like this: INTELSAT, obliged by treaty to engage in globally averaged pricing, will lose its highly profitable dense traffic routes to entrants not so constrained, who will underprice INTELSAT on those routes and ignore the thin traffic routes INTELSAT must serve at a loss.

This is a quite accurate summary of the dilemma faced by INTELSAT. It is not qualitatively different from the arguments used by American communications and transportation carriers when faced with domestic deregulation.

Much has been written about cream-skimming in the regulatory and other literature,[26] but the essence of the problem is easy to state. Cream-skimming is made possible by competitive entry to markets which were previously part of a cross-subsidizing monopoly. Competition, however, forces costs to be aligned with prices in each market; otherwise, the incumbent firm would either be underbid and lose customers (in markets where its prices exceeded costs) or lose money (in markets where its costs exceeded prices).

By interpreting its agreements regarding global and non-discriminatory pricing strictly,[27] INTELSAT has indeed made itself vulnerable to cream-skimming by competitors planning to enter its lower cost, high traffic routes such as the North Atlantic. Preliminary INTELSAT cost studies, at least those during the 1970s, do indeed indicate a subsidy of the Pacific and particularly of the Indian Ocean regions by the Atlantic.[28]

More research is needed if INTELSAT wishes to foreclose the option of "cream-skimming" to potential competitors. More complete and sophisticated cost studies should be conducted to determine whether potential competitors would have the same technology and, therefore, the same cost function, as INTELSAT; and whether INTELSAT's current global tariff structure sets prices above costs in the North Atlantic region. With the continuing rapid evolution of launcher and satellite technology, the answers to these questions may well change every few years.

U.S. Foreign Policy Considerations

Fear of damage from competition to INTELSAT is fed by many concerns other than those that can be expressed in a "technical" (engineering, economic, legal) framework. INTELSAT is regarded by most of its member countries and proponents as the embodiment of an apolitical, nonideological international organization that has harnessed a new technology for the good of mankind. Developing nations have gained access to telecommunications services they could not otherwise afford, and the industrialized world has shared in the technology and aerospace contracting needed to maintain the system. While there were complaints from both European and developing countries about American domination during the early years of INTELSAT, they have become less numerous and strident since the renegotiation of INTELSAT's agreements and with the decline of American usage from over one-half to less than one-fourth of the system. One is hard pressed indeed to find a similarly successful international organization anywhere in the world; there is certainly none providing commercial services on the scale that INTELSAT does.

Global cost averaging was mentioned in the preceding subsection as the potential cause of cream-skimming on the North Atlantic route. Symmetry and fairness demand that positive aspects of this implicit subsidization now be discussed. The net flow of benefits to the (mostly developing) countries of the Pacific and Indian Ocean regions can be seen from one perspective as an extremely successful and unprecedented exercise in multilateral telecommunications development assistance, the kind of "foreign aid" that both donor and recipient nations dream of but seldom achieve through conventional assistance mechanisms, whether bilateral or multilateral. The precise ways in which telecommunications can accelerate economic development—or is itself in part a consequence of such development—are as yet poorly understood.[29] Nevertheless, it is clear that a minimum level of telecommunications infrastructure, including both domestic and international links, is a prerequisite to sustained economic development.

ARGUMENTS FOR COMPETITION IN THE NORTH ATLANTIC COMMERCIAL SATELLITE MARKET

This section examines reasons advanced for allowing competition with INTELSAT on its North Atlantic routes. As with the arguments opposing competition, the arguments below are generalizable to other satellite markets and often to other enterprises or industries.

Diversity of Services and Other Dynamic Considerations
          It has been seen that once multi-product output is considered, traditional conclusions regarding economies of scale and other cost relations had to be modified in a qualitative way. This point applies with particular force when we take so-called dynamic factors into account.

          Most economic reasoning is cast in a static mold for mathematical tractability and ease of theorizing. There are at least two aspects of competition in telecommunications, however, which cannot receive adequate appreciation in a static framework. The first point is that, over time, new services emerge and existing services can become better and more reliable. Thus, analysis based on a fixed set of outputs as the arguments of a cost function cannot do justice to the importance of new services and technologies. The list of new services and techniques in telecommunications is long and varied, including, of course, satellite transmission itself. Second, technological change causes the cost function to shift over time, allowing more output to be obtained from a given set of inputs. These dynamic efficiencies due to changes in the cost function (technology) are to be distinguished from the static cost savings made possible by economies of scale and scope within a given technology.

The burden of economic evidence to date is that these kinds of dynamic efficiencies emerge more naturally and easily in a regime of competition than one of monopoly. This can be seen in the pressure from the business user community, consisting often of multinational corporations as prime movers, to have European PTTs introduce new services vital to conducting international business.[30] User groups and others have argued that national telecommunications monopolies do not have adequate financial incentives to offer such services. In addition, incentives to pursue and adopt more efficient, cost-saving technologies are generally greater in enterprises facing competition or at least having a break-even constraint than in monopolies that can count on taxpayer subsidies to cover their losses.

Contestability of Market Structure

A succinct description of market contestability theory is beyond the scope of this essay.[31] The basic idea, however, is that if markets dominated by a monopolist are relatively easy (inexpensive) to enter and exit, the mere threat, if not necessarily the reality, of entry by rival firms will exert discipline on the incumbent firm to innovate and to price according to cost rather than to earn monopoly profits. Much of the theory of contestable markets centers around the question of how high entry and exit costs are for potential rivals in monopoly markets; what assumptions the incumbent and rival firms make about each other's potential behavior; and what the effects of both entry and the threat of entry are upon the incumbent firm if a market is truly contestable.

In the absence of comprehensive, reliable, and accessible cost studies, we cannot determine whether the North Atlantic is a "contestable" market for commercial public satellite communications. There is some behavioral evidence, however, that it is. Consider first two of INTELSAT's recent service innovations,

Vista service for developing countries and INTELSAT Business Service (IBS) for international business applications.[32] Certainly IBS approximates to some degree the types of services proposed by potential North Atlantic entrants. And the introduction of Vista, along with earlier INTELSAT concessions on domestic transponder leasing and small earth stations, is surely not unrelated to the threat of entry by domestic or regional satellite systems into markets oriented to the needs of developing nations.

There exist data suggesting that telecommunications markets in the United States and the United Kingdom are contestable. The Competitive Carrier proceeding of the FCC, begun in 1979, had the effect of facilitating competitive entry into the U.S. domestic satellite market, making it easier for "non-dominant carriers" to "institute or discontinue service," i.e., to reduce entry and exit costs.[33] The private Mercury consortium was established in the United Kingdom, it seems, precisely to provide a competitive check on the monopolistic power of British Telecom (BT). Although Mercury's incursion into BT's market share has been quite modest to date, evidence suggests that the mere threat of such entry has galvanized British Telecom management to new levels of efficiency and customer awareness.[34]

Optimal Pricing to Maximize Economic Welfare

The discussion of cream-skimming noted that competition has the effect of forcing prices to align themselves with costs market by market and, thereby, following accepted canons of neoclassical economics, to maximize economic welfare. Given what INTELSAT perceives to be an absolute prohibition against relaxing its global pricing policy, competitive entry into its low cost North Atlantic market could indeed inflict "significant" economic harm upon it—harm so extensive, in fact, as to constitute an argument against allowing such competition. Thus, discussion of cream-skimming was included in the arguments above opposing competition. From a broader perspective, however, the creation of welfare-maximizing prices afforded by competitive entry is obviously a consideration in favor of competition.

Measures of global welfare, however, ignore distributional aspects of pricing, as we noted in discussing the development externality features of INTELSAT's globally averaged tariff. Much of the current European coalition opposing deregulation there is composed of groups that would probably become new losers (the poor, the rural, large and well connected contractors) in any redistribution of telecommunications costs and benefits, even though the overall welfare level would increase as a result.[35]

In case a telecommunications monopoly is retained, however, tariff structures are still available that can greatly increase the level of overall economic welfare above that provided either by global averaging or by political compromises.

Stimulation of Overall Traffic by Competition

The claim has been made that in today's overall context of exponentially increasing traffic, competition on the North Atlantic route might actually increase INTELSAT's own traffic, other things being equal. A less extreme version of this argument is that this high rate of traffic growth would quickly compensate any absolute loss suffered by INTELSAT due to competitive encroachments.

Much of this reasoning depends upon assumptions about what market niche is being targeted by competitors. Such markets must include: (1) services that INTELSAT has no intention or capability of offering in any case; (2) services that INTELSAT would offer only if they were offered competitively; (3) services that INTELSAT would offer only if there were a credible threat of their being offered competitively; or (4) services that INTELSAT would offer in any event, regardless of competitive consideration. Naturally, perceptions by INTELSAT and its potential competitors as to what markets would be involved in the case of competition differ widely and are interdependent, and one is tempted to suggest a game-theoretic approach as the most appropriate analytical perspective.[36] In any event, only the actual process of competition, as opposed to a priori reasoning, could determine with certainty how markets, demand, supply, and prices would interact. At this stage the claim that alternative competitive offerings on North Atlantic satellite routes would stimulate rather than depress INTELSAT's traffic, *ceteris paribus*, is speculative indeed.

Competition as an Efficient Discovery Procedure

This final argument offered in favor of competition is general enough to encompass all the rest as special cases. It derives ultimately from the insights of F. A. Hayek, the Nobel laureate economist of the Austrian school.

While closely allied with the neoclassical outlook, Austrian school economists regard the process of competition as uniquely beneficial and efficient, aside from its affects on prices and resource allocation.[37] Hayek considers competition as a decentralized, non-bureaucratic, efficient "discovery procedure" or information system. For example, the best way to determine what market-clearing price and quantity would prevail if competition were to exist would be simply to allow competition to exist, rather than to estimate supply and demand curves and determine where they intersect.

In the context of the North Atlantic satellite market, this argument would go as follows. Neither INTELSAT nor any potential competitors should be excluded from competition on a priori grounds, assuming that routine technical and financial safeguards were enforced. The ensuing competition itself would reveal the underlying cost relationships in the most efficient manner. Several facts, indeed, could be "discovered" by such competition.

First, INTELSAT might not be a natural monopoly at all, given its current level of output. In other words, it might not exhibit cost subadditivity. Diseconomies of large scale organization, for example, might make it possible for INTELSAT and one or more competitors to produce a given output vector more cheaply than could INTELSAT itself. Second, INTELSAT might have a natural monopoly that is sustainable, perhaps by using Ramsey pricing instead of average-cost pricing. In this instance "competition" would consist of the failure of competitors to attain long-term economic viability at prices they must charge to attract customers from INTELSAT; they would ultimately exit the market. Third, INTELSAT might have an unsustainable natural monopoly. Here, as in the first case of no natural monopoly, competitors would be able to enter successfully. Thus, the mere presence of successful competitors would be unable to distinguish between lack of natural monopoly and presence of unsustainable natural monopoly;

additional information, perhaps including cost function estimation, would be necessary. The most important public policy decision would be whether to provide artificial "sustenance" to an unsustainable natural monopoly, for example by prohibiting entry. This would appear to be the theoretical basis of the debate surrounding the economic coordination mechanism of Article 14(d).

Finally, we note that ease of market entry and exit, a prerequisite for market contestability, also contributes to the process of competition as an efficient discovery procedure and information system. The more cheaply rival firms can enter and exit previously monopolized markets, the more quickly information regarding underlying competitive conditions can be "discovered."

CONCLUSIONS

There exist powerful arguments both for and against allowing competition. The strongest reason for prohibiting competition with the global system is the necessity of protecting an unsustainable natural monopoly, if natural monopoly cost conditions (subadditivity) are indeed found to exist. Wise public policy in such a situation would dictate the protection of users' interests by favoring the lowest cost production option, namely a single service provider. This is a static argument which may need to be refined to take into account dynamic factors such as service diversity and stimulation of new technologies. U.S. foreign policy considerations also have a strong appeal but will probably not be decisive.

The strongest pro-competitive arguments cited here include the greater product diversity and attention to users' needs that seem to flourish when competition—or perhaps merely the threat of competition, following contestability theory—is allowed. INTELSAT, however, was seen to be increasingly responsive to special user needs even without actual competiton. Another benefit of competition—its role as an efficient and autonomous discovery procedure and information system—also provides a persuasive case for its adoption under quite general circumstances. Most of the arguments pro and contra that have been assembled here—and indeed most of the historical policy decisions as well—

depend for their validity on certain theoretical propositions that can be corroborated or refuted only by empirical evidence. The empirical element, however, has been conspicuously absent in this essay, because so few comprehensive, reliable, and accessible studies have been undertaken. It seems almost certain that U.S. policy toward competition in international satellite markets will be decided upon in the absence of any such studies. The "technology" for this kind of empirical inquiry—theory, methodology, and data gathering and analysis—exists today as the economist's state of the art. What is lacking is the political will to make resources available to conduct studies of this nature.[38]

It was promised at the outset that no decision about the issue of competition on INTELSAT's North Atlantic route would be made on the reader's behalf, and we trust that this has been the case. Theoretical arguments and historical precedents have been adduced, but empirical studies still remain to be conducted. Will competition itself—which we have not hesitated to characterize as an efficient information system—provide the only information about the consequences of a fateful public policy decision?

NOTES

1. Y. Ito suggests that as technological advances reduce the scarcity of telecommunications capacity, there is a natural evolution from monopoly to regulation. "Telecommunications and Industrial Policies in Japan: Recent Developments," in M. S. Snow, ed., *Telecommunication Regulation and Deregulation: An International Comparison.*

2. M. Kinsley, *Outer Space and Inner Sanctums: Government, Business, and Satellite Communication* (New York: Wiley, 1976).

3. U.S. Department of State, Treaties and Other International Acts Series 5646, "International Telecommunications Satellite Consortium. Agreement Between the United States of America and Other Governments" (Washington, D.C., 1964).

4. U.S. Department of State, Treaties and Other International Acts Series 7532, "International Telecommunications Satellite Organization (INTELSAT). Agreement Between the United States of America and Other Governments, and Operating Agreement" (Washington, D.C., 1971).

5. M. S. Snow, *International Commercial Satellite Communications: Economic and Political Issues of the First Decade of INTELSAT* (New York: Praeger, 1976).

6. R. W. Nelson, "Domestic Satellite Communications: Economic Issues in a Regulated Industry Undergoing Technical Change," in J. N. Pelton and M. S. Snow, eds., *Economic and Policy Problems in Satellite Communications* (New York: Praeger, 1977), pp. 31–61.

7. U.S. Senate, Committee on Commerce, Science, and Transportation, "Long-Range Goals in International Telecommunications and Information: An Outline for United States Policy," 98th Cong., 1st Sess. (Washington, D.C., 1983).

8. M. S. Snow, *International Commercial Satellite Communications.*

9. U.S. Office of the President, "Memorandum for the Honorable Dean Burch, Chairman of the Federal Communications Commission," Washington, D.C., 1970. (Mimeo.)

10. U.S. House of Representatives, Committee on Energy and Commerce, Majority Staff of the Subcommittee on Telecommunications, Consumer Protection, and Finance, "Telecommunications in Transition: The Status of Competition in the Telecommunications Industry," 97th Cong., 1 Sess. (Washington, D.C., 1981).

11. U.S. Senate, 1983. For a complete reproduction of this document with accompanying policy discussion and analysis, see C. H. Sterling, *International Telecommunications and Information Policy* (Washington D.C., Communications Press, 1984).

12. *Ibid.*

13. R. R. Colino, "Statement of Richard R. Colino, Director General-Designate of INTELSAT, Before the Subcommittee on Arms Control, Oceans, International Operations and Environment. U.S. Senate Foreign Relations Committee" (Washington, D.C., GPO, 1983). Mimeo.

14. The most complete single compilation of sustainability and contestability theory is W. J. Baumol, J. C. Panzar, and R. D. Willig, *Contestable Markets and the Theory of Industry Structure* (New York: Harcourt Brace Jovanovich, 1983). An accessible introduction to this subject is in W. W. Sharkey, *The Theory of Natural Monopoly* (New York: Cambridge University Press, 1982).

15. Panzar, and Willig, Baumol, et al., *Contestable Markets.*

16. It is not yet possible to give general criteria for prices that will insure sustainability. Ramsey prices, however, have been shown to assure sustainability under quite broad conditions. See Sharkey, pp. 100–102.

17. Colino, "Statement of Richard R. Colino."

18. The claim has been made (see Colino "Statement," p. 32) that "INTELSAT is not a monopoly" since it does not behave like a monopoly; that is, it does not earn monopoly profits by restricting output, for example. In discussing whether INTELSAT is a natural monopoly we will use a property of the cost function, namely subadditivity, rather than a behavioral criterion.

19. Such global unsustainability is suggested by the statement that "on individual routes or satellites, due to concepts of global interconnectivity . . . the INTELSAT system is highly inefficient." Colino, "Statement," p. 13.

20. M. S. Snow, "Investment Cost Minimization for Communications Satellite Capacity: Refinement and Application of the Chenery-Manne-Srinivasan Model," *Bell Journal of Economics* (1975), 6:621–642.

21. "Statement of Richard R. Colino."

22. E. M. Noam, "Telecommunications Policy on the Two Sides of the Atlantic: Divergence and Outlook," in M. S. Snow, ed., *Marketplace for Telecommunications: Regulation and Deregulation in Industrialized Democracies,* (New York: Longman, 1986).

23. D. M. Lamberton, Australian Regulatory Policy," in M. S. Snow, ed., *Telecommunications Regulation and Deregulation: An International Comparison.*

24. "Statement of Richard R. Colino."

25. E. M. Noam, "Telecommunications Policy on the Two Sides of the Atlantic."

26. A. E. Kahn, *The Economics of Regulation: Principles and Institutions,"* 2 vols. (New York: Wiley, 1970 and 1971).

27. U.S. Department of State, Treaties and Other International Acts Series 7532. "International Telecommunications Satellite Organization (INTELSAT)."

28. M. S. Snow, "Investment Cost Minimization for Communications Satellite Capacity."

29. R. J., Saunders, J. J. Warford, and B. Wellenius, *Telecommunications and Economic Development* (Baltimore, Md.: Johns Hopkins University Press, 1983).

30. E. M. Noam, "Telecommunications Policy on the Two Sides of the Atlantic."

31. Baumol, Panzar, and Willig, eds., *Contestable Markets and the Theory of Industry Structure.*

32. J. N. Pelton, "Communications: Developing Nations Faster," *Satellite Communications* (July 1984), pp. 19–22; and L. Perillan, "The International Switchboard," *Satellite Communications* (August 1984), pp. 27–32.

33. A. D. Lipman "The Rise and Fall of Nineteenth-Century Satellite Regulation," *Satellite Communications* (February 1984), pp. 48–51.

34. C. Jonscher, "Telecommunications Liberalization in the United Kingdom."

35. E. M. Noam, "Telecommunications Policy on the Two Sides of the Atlantic."

36. W. W. Sharkey, *The Theory of Natural Monopoly.*

37. F. A. Hayek, *Der Wettbewerb als Entdeckungsverfahren.* (Kiel, Kieler Vontrage, 1968), n.F.56.

38. Similar frustration has been expressed about the difficulty of attracting funds for international collaborative studies of telecommunications policies. See Snow.

F O U R

INTELSAT: Responding to
New Challenges

JOSEPH N. PELTON

W e live in a changing world. This is particularly so in the world of high technology. In a survey of executives in new start-up venture capital firms, it was found that 43 percent of those surveyed had a computer at their desks, and that these were in use a significant percentage of the time. Let's take another example. It is today possible to transmit a single page of information across the Atlantic Ocean some 50 million times faster than it was a couple of centuries ago. Whichever way one looks—satellites, computers, television, telematics, or robotics—the story is the same: innovation, change, and economic and social revolution, driven by new technologies.

The combination of computer and communications technology, in particular, seems to be giving rise to not only new services and applications in the world of telematics and informatics, but also is giving rise to productivity gains and to new ways of doing business. The banking community, for instance, has strongly moved toward the implementation of telematics technologies for electronic funds transfers. In doing so, they have found that this has eliminated millions of dollars in expenses associated with the "float" delay accompanying funds transfers.

Even more significantly, in economic terms, however, automated funds transfers have led to productivity gains by lessening the time associated with each transaction. Staff time devoted to such exercises have decreased by over 400 percent, and thus led to four-fold productivity gain.

The German Minister of Posts & Telecommunications, in a presentation, recently made an analogy in reverse—between the computer chip and the dinosaur. He noted that the dinosaur, in moving toward ecological extinction, developed an ever larger body with a tiny brain, while the electronic computer chip of today is evolving toward a smaller and smaller body, with an ever increasing ''brain.''

Few people, in light of this evidence, would disagree with the fact that telecommunications is ever more important to the economic and social fabric of all countries the world over—regardless of whether these be developed, industrializing, or developing countries.

Although the importance of communications to industrialized countries is often obvious in many ways, there is increasing evidence (reflected in studies commissioned by the ITU and the Organization of Economic Cooperation and Development) that telecommunications is also essential to the economic progress of developing countries. Cost benefits as high as 100 to 1 have been identified in such diverse parts of the world as Egypt, Kenya, and India. Examples are often as straightforward as the Indian farmer who walked with his ox seven days to pick up fertilizer from a supply depot, only to find the stocks exhausted—returning after a fourteen-day round trip empty-handed. Had he been able to walk even five kilometers to a telephone to make inquiries, most of the wasted effort could have been eliminated. On the island nation of Tonga it was found that export prices, which were negotiated via satellite in the international competitive market-place, increased by as much as 30 percent, while import prices were reduced by a similiar amount as a result of international quotes and bids. The use of affordable telecommunications to establish import and export prices can lead to remarkable differences when contrasted to prices established by the first ship that streams into port on a market day.

INTELSAT: MYTHS AND REALITIES
ABOUT THE GLOBAL SATELLITE COOPERATIVE

Thus, if one accepts the overriding economic and social importance of telecommunications as being clear and well documented, let us now turn to INTELSAT and its particular role in international telecommunications development. In particular, let's focus first on what is INTELSAT? How has it changed the world of global telecommunications at the national, regional, and international levels? And, perhaps most importantly, what changes does it promise for the future? It is impossible to answer these questions without at first clearly understanding what INTELSAT is. This is particularly true because there are many myths and misunderstandings about INTELSAT; how it operates, its organizational structure, what are its goals, or even what the mechanisms are by which INTELSAT's accountability to the world community is maintained.

First, INTELSAT is an intergovernmental international organization, established under two international treaties. The governments of 110 countries currently adhere to the INTELSAT agreements, while 110 designated signatories participate as the working members of INTELSAT. Although INTELSAT is operated on a commercial basis (which means that all members must pay for services received), it is also a nonprofit cooperative, and services are made available to all countries of the world on an open and nondiscriminatory basis. Thus, while INTELSAT has a membership of 110 countries, it actually provides services to 170 different countries and territories around the world, including countries that are democracies, planned economies, monarchies, or that have any other form of government.

There have also been attempts at times to characterize INTELSAT as a "typical" international organization—as being large, bureaucratic, and not innovative. These unfortunate characterizations are certainly inaccurate in the case of INTELSAT. It has a small staff (under 600) that operates the global satellite system, with only 30 percent of revenues devoted to operating cost. INTELSAT has achieved remarkable breakthroughs in cost efficiencies. The cost of INTELSAT's communications capacity in orbit per year has fallen by almost two orders of magnitude since IN-TELSAT began operations in 1965.

There have also been attempts in the same respect to characterize INTELSAT as some sort of an international or multinational monopoly that arbitrarily controls the international communications marketplace. In this view of INTELSAT, it can, like a monopoly, maintain rates at very high levels and make larger profits. Again, this characterization is almost the exact reverse of the actual situation.

First, INTELSAT's rates are remarkably low and will be even lower in the future. Since INTELSAT began operations in 1965 it has reduced its rates on twelve different occasions. Furthermore, if one adjusts for inflation, the cost for INTELSAT service today is almost one-twentieth (or 5 percent) of the charges that applied when operations began with Early Bird, in 1965. Indeed, INTELSAT has done such a good job of reducing rates for all of its users—large, medium, and small—that today its rates reflect only a very small proportion of what the end-user pays and, in fact, are typically 8 percent or less of the amount paid for a complete international or data circuit.

INTELSAT, indeed, does not have a monopoly on international telecommunications. Since the beginning, INTELSAT has had, and continues to have, serious competition from submarine cables—most recently in the form of highly cost-efficient fiber optics cable systems. In addition, certain regional systems which were evisioned within the framework of the INTELSAT agreements have been successfully coordinated with INTELSAT, with regard to services provided within definable regional areas of the world, as reflected in the coordinations of the Arabsat, Eutelsat, and Southeast Asian Palapa systems.

It has also been maintained that the INTELSAT structure is established so that only monopoly PTT organizations can utilize the INTELSAT facilities and, thus, it serves to stifle competition or deregulation at the national level. Again, experience in both the United Kingdom and the United States has demonstrated that this also is an inaccurate characterization of the IN-TELSAT structure. The United Kingdom's government has authorized establishment of two organizations—namely British Telcom

International and Mercury—to access the INTELSAT system and to provide services to end users in the United Kingdom. In the United States, the FCC has authorized international carriers who wish to provide either video services or digital business communications services from customer premise or small earth terminals serving as urban gateways to own and operate such earth stations and to obtain the services from INTELSAT on a "de-bundled" rate basis, through the U.S. signatory.

These changes at the national level, reflecting goals to increase competition and deregulation in those countries, have been accommodated with the INTELSAT system. In many ways the INTELSAT system should perhaps be seen as equivalent to an international railway, upon which countries or commercial organizations can place their trains or boxcars, or even containerized packages, of information which can be transported on a global basis. There are today, in effect, transoceanic satellite and cable "telecomunications railways," as represented by the INTELSAT system and by submarine cable systems. The question is, how many railways should be built before serious overinvestment occurs?

In this respect, it was recently pointed out to me, by the chief executive of Telecom Ireland, that the problems posed by a totally unregulated and competitive market in the telecommunications areas (as represented by the United States' approach to international telecommunications) are both potential overinvestment, and attendant great risk of bankruptcy, which has many repercussions. The worst repercussion, as seen by my colleague from Ireland, was the likelihood of a very heavy drain of capital away from Europe and third world countries (where capital investments are desperately and urgently needed) into the U.S. market. This "unbalanced" regulatory approach, which skews investments not only at the national level but at the global level as well, is conducted without any reference to social need or redundancy of investment. This, I believe, is a serious point for discussion. Should there be regulatory distinctions between national and international markets? And if developing countries cannot compete for capital investment successfully, what recourse do they have?

Overcapitalized telecommunications investment in the United States, in my view could, in the next decade, be among one of the contributing factors in developing countries' not being able to finance and capitalize needed new telecommunications projects. My Irish friend felt this could particularly be so in light of the arbitrarily high interest rates being maintained in the U.S. market. The process by which bankruptcy is the only instrument of accountability in capital investment decisions is thus perceived as grievously indifferent to global telecommunications needs, particularly in the third world countries.

INTELSAT's ACCOUNTABILITY

This leads us to the issue of INTELSAT's own accountability. It has been stated, for instance, by critics of INTELSAT, that it is accountable to no one. This, of course, is demonstrably not the case, but I believe it is important to explore and examine those instruments of accountability that do exist.

Competition

INTELSAT, as previously noted, experiences competition with fiber optics systems. The fact that the competition is at work would seem to be strongly indicated by the fact that INTELSAT's space segment capacity, as measured in megahertz of capacity in orbit per year of operation, is today close to 100 times more cost efficient than when service began in 1965. Furthermore, INTELSAT's rate reductions (which have already been noted) over the last nineteen years are among the most dramatic of any service offerings provided in the world, and are perhaps surpassed only by the computer industry. It should be noted that, in an industry in which there are both economies of scale and economies of scope, unlimited competition is not required to produce the highest form of cost efficiency. In fact, all that is needed is sufficient competition to achieve the balance between competitive pressures and the maintenance of economies of scale and scope.

Arbitration Procedures

Built into the INTELSAT agreements are provisions for arbitration, which allow any country or signatory who disagrees with a major INTELSAT decision to seek redress through arbitration proceedings. It is a great testament to the political efficiency and the objectivity of INTELSAT's decision making process that no country in the history of INTELSAT has ever invoked or utilized the arbitration proceedings.

It is also perhaps significant to note that, despite elaborate procedures that exist within INTELSAT for voting upon issues where and when necessary, actually resorting to a vote is a rare exception and, indeed, 95 percent of all decisions in the various bodies of INTELSAT are taken by consensus. This, again, is largely a result of having objective measures upon which the merits of decisions can be assessed and consensus achieved. Thus, INTELSAT is dramatically different from the "bloc voting patterns" and political decision making processes represented by the UN. It is significant to note that the mechanisms to protect objective decision making exist and, most pointedly, these exist in the arbitration provisions of the INTELSAT Agreement.

Checks and Balances

Another major element of accountability is the built-in checks and balances of the INTELSAT organization. There is, within the INTELSAT organizational structure: (1) a Board of Governors with a weighted vote; (2) a meeting of signatories, which gives all countries and their signatories, who act as owners, a chance to make major policy decisions as well as those involving capitalization limits. The meeting of signatories operates on the basis of one country, one vote; and (3) finally, there is an Assembly of Parties (which involves only governmental entities). This body also takes major policy decisions concerning INTELSAT, including coordination of other satellite systems and the amendment of the INTELSAT agreements themselves—again, on a one vote per country basis.

This decision making structure thus has built into it significant checks and balances: to allow governments to have checks upon signatories; for all signatories to have checks upon the signatories with the largest investment shares of INTELSAT; and, indeed, for the largest signatories (with the greatest investment in INTELSAT) as represented on the board, to have some check upon the wishes and desires of the overall membership who might conceivably want to pursue the programs and activities that could be against the best financial interests of those who have invested the greatest amount of capital.

Right to Amend the INTELSAT Agreement or Operating Agreement
Any country that believes that the INTELSAT structure, its decision making process, its capitalization procedures, its charging procedures, or, for that matter, even approaches to competition, are at variance with the needs of the current international communications and information marketplace and the broader world community has the right to initiate an effort to renegotiate the INTELSAT agreements. Indeed INTELSAT, which was established in 1964, has already been through a process (1969 to 1971) to amend the INTELSAT agreements to reflect a new international consensus on how INTELSAT should be structured, how it should operate, how it should charge for service, how it should be capitalized, and other such key issues. It is perhaps somewhat ironic, in this respect, that the country which has the largest say in the initial structure of INTELSAT, and again in its restructuring during the 1969 to 1971 negotiations, is today the country that sees the need for further significant changes to the INTELSAT Agreement.

INTELSAT AND THE U.S. POLICY OF DEREGULATION OF TELECOMMUNICATIONS: CONFLICT OR COMPATIBILITY?

The other members of INTELSAT, of course, respect the sovereign right of the United States to seek to move INTELSAT in new directions. The concern is that the United States has shown some inclination to move unilaterally to institute change, without international negotiation, and outside of the procedures established by the INTELSAT agreements. This is not a minor issue. More than seventy countries have placed letters and diplomatic notes on file with various agencies of the U.S. government with regard to their concerns about "unilateral" approval of so-called "private" satellite systems. Many of these countries have very clearly stated that if the United States wishes to change the structure of INTELSAT it should do so through the authorized procedures, and not attempt to reinterpret independently the INTELSAT Agreement so as to achieve a restructuring of INTELSAT on a de facto basis.

Recent studies of international U.S. trade policies have shown that previous attempts by the United States to redefine unilaterally multilateral agreements in the transportation field have met with mixed success and, even at times, outright failure and embarrassment to U.S. policymakers. There is no particular reason to suspect that similar uncoordinated and unilateral initiatives in the telecommunications area might not lead to similar results.

INTELSAT, in my view, is not only highly accountable to the international communications and information marketplace, but receptive to constructive proposals for change and innovation. The mechanism that has produced accountability—namely, the INTELSAT agreements—has also allowed INTELSAT to be an effective global common denominator, a bridge between and among all of the various countries of the world, regardless of their levels of economic development; regardless of the sophistication of the telecommunications and information infrastructure; and regardless even of whether they are members of INTELSAT or not. In this respect, INTELSAT is strikingly unique among other international organizations which, unlike INTELSAT, have frequently been marred by serious political bickering and a decision making process that is often characterized by politics first, and objective decision making on merits last.

There are many who fear that U.S. government policy issues could not only serve to greatly politicize the INTELSAT organization, but ultimately lead to its demise as an effective international institution. This, I think unlikely. I do feel, however, that it would be a disservice not to underline and emphasize the magnitude of the risk that could be involved if the United States were to proceed to change the nature of the INTELSAT system on a unilateral, de facto basis, rather than to seek formally a new set of rules to reflect a new global consensus.

First, it should be noted, for the sake of clarity, that there are no restrictions that limit the ability of the United States to encourage and to achieve new and effective means of deregulation and pro-competitive policies concerning the use of access modes to the INTELSAT system, as long as these were executed on a strictly domestic basis under the regulatory authority of the FCC and consistent with the Communications Act of 1934 and the Comsat Act of 1964, as both have been amended. Indeed, U.S. policy decisions by the FCC have not only been accommodated by INTELSAT itself, in terms of allowing a large number of new U.S. international carriers to access the INTELSAT satellite system for video and digital communications, but are being accommodated in other parts of the world, in terms of either signed new operating agreements or letters of intent to operate with new U.S. entities. Such letters or agreements have been signed in the United Kingdom, West Germany, Switzerland, and elsewhere.

This shift toward international "service" competition in international telecommunications, plus increasingly sophisticated and earnest competition between INTELSAT and fiber optic cable systems, could, without any further facility competition, fundamentally change the focus and duration of international telecommunications. In this respect, U.S. policymakers need to consider what objectives have been or will be achieved under changes now approved and what are the "pros" and "cons" of pushing beyond the scope of changes already made.

INTELSAT AND INNOVATION FOR THE FUTURE

The ultimate success of INTELSAT, of course, will not hinge on the number of mechanisms available to achieve accountability or the protections provided in the INTELSAT agreements. The ultimate test will be, in fact, the international communications and information marketplace.

In short, will INTELSAT be able to expand the volume, scope, and flexibility of its service offerings to effectively meet new emerging demands? Also, can and will INTELSAT keep users happy? In this respect, INTELSAT's record, by objective measures, would appear to be extremely impressive. The INTELSAT system has gone from 0 percent of the international overseas transoceanic telecommunications market to approximately two-thirds of global traffic demand in this area.

INTELSAT has also become the predominant supplier of international video relays on a transoceanic basis, even though new wideband fiber optic systems should be able to provide strong and effective competition in this area. INTELSAT, for instance, in anticipation of future market demands, has recently approved and introduced a digital television service which, within the next five years, should allow the provision of video services at significant rate reductions. In the meantime, different priority levels for video services are allowing cost reductions now. Equally important, television services can now be leased on a full-time basis for different time periods, ranging from one to seven years.

Also for low volume users, there are now part-time lease services, plus peak and off-peak occasional use rates that allow users to tailor their distribution services to their specific needs. Digital signal processing of the future will allow multiple TV channels to be sent through a single transponder. Into the bargain, there can also be a parallel reduction in the size and cost of earth stations that will receive such digital services.

INTELSAT has greatly diversified its services over the last twenty years in all areas—not just video services. In response to user needs INTELSAT has introduced such innovative new services as leased domestic telecommunications services (now in twenty-five countries); provision of maritime mobile services (leased to INMARSAT); the provision of new business digital communications services to customer-premise earth stations, a service known as the "INTELSAT Business Service"; and even a highly sophisticated new data distribution service, called "INTELNET," which can provide data links through microterminals as small as two feet (or 65 centimeters) in diameter.

On the horizon, INTELSAT expects to introduce, in the near future, electronic document distribution services. This might ultimately lead to highly interactive INTELNET data broadcast and distribution networks. Also by the 1990s, INTELSAT will likely be providing high definition television services. INTELSAT will also likely move even further toward diversification of a tariffing structure to allow tailored telecommunications and broadcasting needs to respond to new market demands.

Although it is in many ways clear and reasonable to compare INTELSAT's technological and service innovations record with that of domestic satellite systems, such as exist in the United States, Canada, Japan, Europe and, indeed, a number of such developing countries as India, Indonesia, Mexico, and Brazil, one area of significant difference between INTELSAT and such other systems should be particularly highlighted. INTELSAT, more than any other system in the world, has as its objective the provision of global interconnectivity. At times, much attention is directed to the tradeoff between the use of radio frequencies and power levels, in order to achieve a maximum amount of capacity. IN-TELSAT must, however, design its satellites and its services on the basis of a three-way trade among radio frequencies, power levels, and interconnectivity. Global interconnectivity, particularly for low density traffic routes, does not come cheaply or easily.

It is the INTELSAT objective of achieving global inter-connectivity that forces the INTELSAT space segment design, in terms of its use of frequency and power, to be less cost effective than domestic systems. It is, in fact, only due to such aspects as lifetime extensions, economies of scale and scope, that INTELSAT transponder costs per year in orbit have been able to be maintained in surprisingly close proximity to domestic systems. The INTELSAT system, for instance, provides on a global basis some 1,500 earth station-to-earth station pathways. It is significant to note in this respect that half of these pathways (that is, in excess of 750 of them) provide INTELSAT with less than 10 percent of its revenues. Furthermore, it is equally significant that about 10 percent of the pathways represent approximately 50 percent of INTELSAT's revenues.

Thus, if one were to point to a single characteristic that is unique, special, and fundamental to the INTELSAT global system design, it would be this aspect of serving as the global common denominator that links developing countries, newly industrializing countries, planned economy states, and highly advanced countries together into a vast network that is the INTELSAT global system. On the average, each INTELSAT satellite in operation provides 100 international pathway links. No other satellite system comes close to achieving this type of global interconnectivity.

This aspect of INTELSAT is important in another way—in the conservation of the use of the orbital arc. INTELSAT's seventeen satellites from only seventeen orbital locations serve 170 different countries and territories for international services. They also provide eight international television networks, twenty-five domestic satellite systems, and an important element of mobile services to the world maritime community. In an era of satellite proliferation, the world's largest common user satellite system is the most effective conservator of the geosynchronous orbital arc.

It is the future potential of the INTELSAT Business Service to achieve a multipoint-to-multipoint network, linking ultimately thousands or even tens of thousands of points, that represents in many ways the greatest potential of INTELSAT to compete effectively with the fiber optics cable systems of the future. Certainly, those who feel INTELSAT needs to be stimulated to greater heights of innovation and market responsiveness should not doubt that these forces exist.

Not only are there fiber optics systems and new digital processing techniques that will serve to push the cost of INTELSAT's services down, but there are the new integrated services digital networks standards (ISDN) that should aid in maintaining the quality and integrity of future telecommunications services. The policy of stimulating, multiple and diverse telecommunications systems, using dozens of terrestrial and space telecommunications technologies, and hundreds (or even thousands) of networks, will make standards and high quality service increasingly difficult to maintain. This rather basic and fundamental conflict has been conveniently swept under the carpet, but it won't go away. The AT&T divestiture decision alone will make intersystem network compatibility a major operational, technical, and standards issue for decades to come in the United States; nor will the rest of the world be insulated from the controversy.

The need to provide effective interconnection to fiber optics systems and domestic and regional satellite systems in the ISDN mode of operation will be, indeed, one of the great technical challenges of the 1980s and 1990s. It is, in many ways, remarkable, to me at least, that the INTELSAT system has been able to introduce a very high rate of technological innovation and that it continues to diversify service offerings responsive to the needs of very highly sophisticated users (such as banks, oil companies, and other multinational enterprises), while at the same time continuing to be highly responsive to the needs of third world and developing countries.

In this respect, INTELSAT has introduced, within the last few years, VISTA low density, thin route communications service, the INTELSAT Assistance and Development Program (IADP), and, during 1985 and 1986, it will be conducting Project SHARE (a test and demonstration program related to health and rural education). We have also initiated a serious study of what we call the INTELSAT Development Fund which, if established, will help in the financing, as well as in the design, of telecommunications systems in rural parts of the world, with such financing covering not only the ground segment, but also terrestrial interconnect and terminal equipment as well.

It will likely be one of INTELSAT's greatest challenges in the next decade, to be able to design space segment that remains, on one hand, extremely cost effective and responsive to customer demands but, at the same time, achieves global interconnectivity and responds to the needs of countries at all levels of economic development. In this respect, techniques such as cross-strapping of frequencies, on-board processing of satellite signals, electronic hopping beams, and even intersatellite links, may be essential to INTELSAT's meeting its multiple missions in the 1990s.

INTELSAT AS A VIABLE CONCEPT FOR THE FUTURE: PRIVATE TELECOMMUNICATIONS CARRIERS, NATIONAL PTT ENTITIES AND THE GLOBAL SATELLITE SYSTEM: HOW DO THEY RELATE TO ONE ANOTHER?

I would like to close by presenting a very brief comparative analysis of government-controlled PTT entities, on the one hand, versus private enterprise (market-driven, competitive, and deregulated), on the other. It is often assumed that monopolies and government-controlled enterprises can best achieve such goals and objectives as universal access to all users, and the provision of subsidies for rural and isolated customers in the provision of such basic and traditional telecommunications services as switched telephony. It is also widely assumed, however, that such entities may well tend to maintain rates at higher levels than are necessary; that the organizational structure of such institutions are very bureaucratic and slow to respond; and that they do not provide innovative and new services in a timely manner.

On the other hand, it is often assumed that private, unregulated, market-driven organizations are quick to respond to service innovation, will depreciate obsolete equipment rapidly and introduce new facilities or services at the earliest possible date, and will be highly responsive to very sophisticated communications users who demand innovative services, flexibility and service offerings, and the lowest possible tariffs.

Certainly, as is the case with many stereotypes, there may well be both truths and errors in such attributions. It is important for serious policymakers in the field of international telecommunications to look beyond stereotypes to understand when such attributions are correct and when they are incorrect. Certainly, I would argue that the INTELSAT framework was carefully and extremely wisely drafted, and that it, in many ways, contains a beneficial mix of attributes.

INTELSAT has sufficient competition to innovate, introduce new technologies, and develop new services quickly. It is a nonprofit cooperative. It does not have a profit motive nor a "subsidy" requirement in a classic economic sense, to retain prices at high levels. INTELSAT cannot retain excess revenues under the INTELSAT Agreement, so again it only has motivation to grow, expand, and reduce the cost of its services. INTELSAT does not give special breaks in charges or services to any single set of users, because this is prohibited under Article 5 of the INTELSAT Agreement. Therefore, all users—big, medium, and small—know they are being treated fairly and equitably. The cumulative effect of worldwide participation provides sufficient traffic volume to keep prices low for everyone.

Finally, the flexibility and responsiveness of INTEL-SAT's management is shown in the satellite system's service innovations, reliability, and low cost of service. It is also shown in small staff size, use of contractors for most major procurements, and in the nonpolitical recruitment on a global basis of the best staff. All in all, INTELSAT is a remarkable and unique organization that does not compare at all with the analytic framework established by many national policymakers when they try to view INTELSAT from the perspective of a deregulated commercial enterprise or governmental monopoly. INTELSAT is neither, and it should be assessed and analyzed on its own very special merits.

In many ways, INTELSAT has indeed produced the best of all possible worlds. Furthermore, governments, at the national level, have a tremendous amount of flexibility in regard to the form, nature, and characteristics of their national participation in the INTELSAT system, so as to achieve the best balance and mix of characteristics between a competitive commercial enterprise and governmental telecommunications enterprise, whichever they feel is most appropriate to their national needs. The country that wishes to encourage maximum competition and competitive access, as well as the introduction of innovative services, can easily do so. Furthermore, another country, more concerned with the establishment of universal access; the establishment, on a national basis, of basic telecommunications services; or the implementation, in the telecommunications field, of social services (such as health and education), can use the INTELSAT system to achieve these goals and objectives as well.

CONCLUSIONS

In short, INTELSAT is a unique twentieth-century mechanism. It seeks to combine rather special and valuable characteristics, well-suited for high technology commercial ventures requiring international collaboration, compatibility, and common capital investments. Such strengths that INTELSAT possesses, particularly in the form of effective north-south political, economic, and technical cooperation, should be built upon and improved in the 1980s

F I V E

The Reality of Change, Satellite Technology, Economics, and Institutional Resistance

CHRISTOPHER J. VIZAS, II

Thirty-five years ago, the visionary author Authur C. Clarke conceived a new idea in telecommunications—a communications satellite in geostationary orbit. A little over a decade later, the government of the United States set out to make the idea a reality. By the mid 1960s, the idea had been translated into a fledgling industry built around a global commercial consortium. In the 1970s that worldwide consortium was augmented by a number of national and regional systems. As the 1980s begin more new systems are proposed each year. Indeed, the business of satellite communications seems to be on a path of accelerating change. That change is extraordinary given the already revolutionary developments in the technology and economics of the satellite industry in its first twenty-five years. But the essence of satellite communications is, and will continue to be, change. The changes in technology over the last quarter century (and the changes in the economics of satellite systems that followed) have brought the world faster, less expensive, and more efficient communications.

Those changes have been brought about in part by the initial research and development program undertaken by the United States government, in part by the research efforts of the manufacturers of the satellites and the earth stations, to some degree by the research efforts of the satellite operators, but in critical part (and increasingly in recent years) by the demands and needs of the end users of the services that satellite communications offer. These combined changes in technology and economics have lowered both the complexity and cost of using satellite communications—most of the national and regional systems have been key instruments, as well as beneficiaries, of these developments. This reality of change has permitted satellite communications to remain a business of dreams—one can be reasonably certain that the technology and economic arrangements of tomorrow will provide more effective and efficient services than those that exist today. The only threat to the dreams will come from attempts to hinder change.

Unfortunately, the administrative and commercial arrangements created only a quarter century ago as the underpinnings for the nascent satellite communications industry may prove to be barriers to the kind of innovation and change that they were intended to nurture. As happens all too often with human institutions, the institutions become bureaucratized and, ultimately, dysfunctional; they lose sight of their objectives, of serving the needs they were created to serve in the most effective way possible. While it is not yet clear that the existing structure of satellite communications, particularly the INTELSAT organization, has become too rigid to adapt to change, the danger clearly exists.

To understand this danger, and to appreciate the kinds of flexible arrangements that may be necessary in the next few years, one must review the origins and structure of INTELSAT, the limits imposed by the 1972 Agreement, and INTELSAT's recent reactions to alternative specialized satellite communications systems.

THE ORIGINS OF INTELSAT

The creation of INTELSAT was a direct outgrowth of the actions of the United States government in the Communications Satellite Act of 1962. In creating Comsat, the President and the Congress mandated the establishment of a global commercial satellite communications system in which Comsat would be the U.S. participant.

After the embarrassment of the first Soviet Sputnik launch, the United States adopted a policy of demonstrating its technological capabilities in space as an urgent priority. There was no time to await technical developments which could ensure a multiplicity of competing systems. With the imminent threat of the introduction of a Soviet space communications capability, the goal of the United States was to establish a Western commercial system first. Thus, the Congress mandated Comsat's role in a single satellite system, not necessarily as the best way to proceed, but as the most effective and immediate way to proceed, given the technical, economic, and political constraints of the era.

Indeed, there was considerable concern that competition be preserved to the extent possible. In his July 24, 1961, speech on space communications, President John F. Kennedy stated that private ownership of the global system was favored with an assurance of "maximum possible competition." The drafters of the legislation, therefore, provided for competition where feasible, such as in procurement of system components.

Aware of the potential limitations of the global system of which Comsat was a part, the legislative history of the Satellite Act also shows that, in authorizing the system that became INTELSAT to operate on a common carrier basis, Congress recognized that the proposed structure might not be optimal for all future demands for satellite communications, and that the door should be left open for alternative technical and entrepreneurial advances. Congress enacted Section 102(d) of the Satellite Act to ensure that the FCC would have the power to authorize future alternative satellite systems, if warranted. The importance of this was emphasized by Senator Church when he stated "we cannot now foretell how well the corporate instrumentality established by this act will serve the needs of our people . . . [C]ertainly this enabling legislation should not preclude the establishment of alternative systems, whether under private or public management."

INTELSAT STRUCTURE

Following the passage of the Satellite Act and the creation of Comsat, preliminary discussions with other governments were conducted and INTELSAT was created in 1964. Established initially under an "Interim Agreement," the arrangements between governments were completed and a definitive agreement established in 1972. INTELSAT is owned and governed by signatories appointed by its 110 member nations. Signatories are invariably the communications operators of a nation; they range from government agencies, to government owned monopolies, to companies with a mixture of government and private ownership, to purely private enterprises. For example, the signatory designated by the United States through statute is a private company, the Communications Satellite Corporation ("Comsat").

In contrast, the signatory designated by the United Kingdom, British Telecom, is a corporation jointly owned by the government and private investors. Indeed, the Thatcher government apparently intends to sell additional substantial portions of the government equity interest to private shareholders. For France, the signatory is the government of France, represented by the Ministère des Postes, de Télécommunications et de la Télédiffusion.

Under the agreement that governs INTELSAT, the day-to-day operations are carried out by an "Executive Organ" of several hundred employees headed by a director general. Executive decisions and policy direction are provided by a Board of Governors dominated by the major investors (voting is weighted according to ownership interest). The broad policy judgments and long-term direction are set by an Assembly of Parties composed of representatives of the member governments. The representatives at the Assembly are often not the signatories but officials of the governments. Put bluntly, INTELSAT is not so much an organization of governments as a consortium of national communications monopolies operating for their commercial advantage.

DEVELOPMENT OF INTELSAT

As noted earlier, when INTELSAT was established it was conceived as a single global system largely as a result of a combination of technical, economic, and timing constraints. In the last two decades, the timing consideration has entirely disappeared and the technology and its economics have changed radically. Technological advances permit continually more efficient use of the radio frequency spectrum and the orbital arc. Advances in the design of satellite and earth station hardware also have resulted in reduced costs for satellites, for earth stations, and, most important, for entire systems. Where once only a limited number of general purpose communications satellites were technically and economically feasible, today's technology and the consequent economics permit an increasing number of systems tailored both to particular geographic areas and to particular user groups.

Thus, INTELSAT, while emerging as the dominant international communications satellite provider, is by no means a global monopoly. The first departure from the single global system was the development of domestic communications satellite systems separate from INTELSAT. While INTELSAT provides some facilities for domestic use, most domestic satellite communications are provided by national carriers. Domestic satellite systems exist, or are planned, in a number of INTELSAT member nations; systems in Brazil, Canada, France, Indonesia, Japan, Luxembourg, the United States, and the United Kingdom are examples. Indeed, the highly competitive U.S. communications satellite industry has, in aggregate, as much transmission capacity as INTELSAT. Non-INTELSAT domestic satellite communications facilities have existed for almost a decade.

More recently, INTELSAT's international dominance has been eroded and more erosion is planned. The transborder use of domestic satellites in North America, the Palapa system, Arabsat, Eutelsat, and a recent proposal by Luxembourg represent regional satellite communications systems which serve expanding international as well as domestic markets in their particular parts of the world.

Only in transoceanic satellite communications has IN-TELSAT maintained the dominant role; even there, however, the consortium is being challenged. Both Unisat in the United Kingdom and the French Telecom were designed with the capability to provide transatlantic facilities. Unisat offered its transatlantic capacity on a leased basis to INTELSAT, which rejected it. Telecom has been followed in France by the mysterious Videosat proposal. Japan published a study last year of the practicability of a non-INTELSAT transpacific satellite communications system. Finally, in the United States, five private companies have filed for permission to provide intercontinental satellite communications facilities: four across the Atlantic and one between North and South America. Three of the U.S. proposals seek to provide specialized Customer Premises Services (CPS) while the other two seek direct competition with INTELSAT in public switched services. Orion Satellite Corporation, which made the first of these proposals, sought only CPS and it appears that that is the kind of arrangement that the United States government may permit.

SUCCESS OF INTELSAT

INTELSAT has been a success, although not an unqualified one. It is one of the dominant communications satellite service providers in the world and the dominant service provider for international communications. Given its size, its market position, and its unique relationship with national telecommunications operations around the globe, it will remain not only viable, but dominant in intercontinental satellite communications. In short, INTELSAT has achieved many of the objectives set for it in 1964; it is a resounding commercial success.

INTELSAT has not achieved all of its objectives, however; it still does not provide direct and cost-effective public network services to many parts of the developing and newly industrialized world. The INTELSAT system, largely because of its design, doesn't meet as efficiently as possible the technical and economic needs of many geographic areas—strategically important areas such as the Pacific Basin, for example. But INTELSAT has the wherewithal to achieve this critical objective; if it adheres to its mandate under the INTELSAT Agreement.

INTELSAT was never intended to do everything. It was primarily intended to provide facilities and services for switched voice and equivalent data services and to provide capacity for conventional television transmission.

Article 3 of the INTELSAT Agreement clearly recognizes that intention. Article 3 was written to assure that INTELSAT would pay attention to its primary purpose and, at several points, admonishes INTELSAT's management to ensure that their plans do not impair that primary purpose.

Throughout INTELSAT's two decades of life, the less powerful member nations have been promised by the INTELSAT Board of Governors and, more recently, the Executive Organ, that INTELSAT will meet the objective for which it was created and focus its resources on its prime mission. Thus far, INTELSAT has made substantial progress toward achieving a low cost, worldwide public network; with the capital resources that will be generated by INTELSAT's revenues over the next decade, INTELSAT could finally fulfill its promise and accomplish the mission for which it was created.

LIMITS OF INTELSAT

Today, however, much of the INTELSAT promise is still rhetoric and not reality. Many member nations, in large part because of constraints on space segment technology (most of which are rapidly disappearing), have been unable to take their place as full partners in INTELSAT.

INTELSAT chooses to employ satellite space segment technology designed to serve high volume, public switched traffic best, particularly in the North Atlantic market. As a result, earth station investment (not to mention operating costs) averages twice as much per half circuit for developing and newly industrializing nations as it does in industrialized nations. A different techno-logical choice by INTELSAT might raise space segment cost some-what, but could sharply reduce earth station cost, particularly for small users, thus creating substantially lower total system costs for developing nations. One example of how INTELSAT might achieve such an impact would be through separate satellite designs for each ocean region which employed spotbeams covering gate-ways within each nation. Rural and domestic services could be most efficiently and cost effectively served by regional or national systems (offered by INTELSAT and others) which provided higher power levels and greater connectivity within national and regional markets. The major partners in INTELSAT, however, have chosen to ignore such design alternatives, leaving the developing world to carry a much heavier burden of ground segment cost than is necessary.

A fairer allocation of INTELSAT's resources could alter this, assuring real access and the best possible satellite commu-nications to all who want to use the INTELSAT space segment.

Diversion of resources to the construction (in some cases, the reconstruction) of facilities for uses that are beyond INTELSAT's primary mission will constrain INTELSAT's ability to make sure that all its member nations become full members in operational terms. Equally important, it is unwise and inappro-priate under the terms of the agreement for INTELSAT to divert resources to monopoly communications markets over which it has no monopoly mandate and which may be better served by alternative facilities arrangements. INTELSAT was never intended to do everything. It was primarily intended to provide facilities and services for a global public network. Article 3 of the INTELSAT Agreement clearly recognizes that intention.

And Article 3, while permitting INTELSAT to provide domestic satellite services and specialized services, does not give INTELSAT any monopoly over such offerings and only permits them to be provided if they do not damage the primary mission. As the European Space Agency (ESA) has noted, the Agreement, particularly the wording of Article 14, is the result of a difficult compromise reached, in the comparatively fast moving world of space, a long time ago. The text is purposely cautious and diplomatic. In fact, according to the ESA, the spirit of Article 14 was to adapt to future circumstances in a pragmatic fashion. Combined with Article 3, it protects against unreasonable moves either by INTELSAT or its members. No one's interests are served by interpreting the Agreement too narrowly, whether to give INTELSAT too much reach or any one member too much leeway. Rather, all interests are served by understanding that INTELSAT cannot cover the entire universe of communications needs and was never intended to do so. While protecting and preserving INTELSAT's primary mission as a provider of public telecommunications services is vital, equally vital is assuring that all the needs of commerce and society are met most effectively. The structure of telecommunications must reflect its role as a resource for other and usually more important social and economic activity.

DANGERS OF INTELSAT UNWILLINGNESS TO CHANGE

INTELSAT's recent attacks on alternative systems which employ specialized technology (e.g., technology tailored to particular uses, such as CPS in a particular geographic region), coupled with its unwillingness to move to satellite designs intended to lower total system costs for the smaller users, threatens to create conflicts and barriers to the entry of new systems and services. Out of those conflicts and barriers can only grow real dangers to fulfilling the promise of satellite communications, and chief among the dangers are: (1) that INTELSAT, by attempting to be all things to all people and diverting resources from fulfilling its primary mission, will damage its ability to provide the global public service network which it is charged with providing; and (2) that other vital needs of international commerce will remain unmet, or that the ability to meet them will be delayed, damaging economic activity vital for prosperity and growth in the industrialized and developing world alike.

As explored earlier, INTELSAT has been a success, but not an unqualified one. One result of this less than complete success has been the emergence of regional alternatives which fulfill public telecommunication needs that INTELSAT has not met, regional alternatives which, in many circumstances, provide less expensive service on a full-system cost basis. Indeed, a major reason for INTELSAT's inability to meet the basic communications needs of the developing world has been its failure to develop economically efficient satellite resources which could be accessed by low cost earth stations. The technology exists to permit IN-TELSAT to provide efficient satellite resources both for so-called high volume routes, such as those between the United States and Japan, and for low volume routes, such as those between island nations in Micronesia and other areas of the Pacific Basin. That technology forms the technical basis for proposals such as Japan's for the Pacific Basin or that of National Exchange, Inc., for the domestic U.S. market. To move from its current system design to one more responsive to the needs of the developing world will take redesign and the reapplication of resources by INTELSAT, such as the redesign recently suggested by Mr. Chitre, Director of Systems Planning for INTELSAT. That INTELSAT has not moved more rapidly in this direction is unfortunate. The failure to move as rapidly as possible encourages the growth of possibly unnecessary alternative public telecommunications services.

In short, INTELSAT should be applying its resources to assure quality international public communications services to all its member nations. The diversion of resources to other needs both detracts from its primary mission and is highly impolitic. Indeed, given the strictures of Article 3 of the INTELSAT Agreement, it would appear to be improper for INTELSAT to pour precious resources into serving the specialized needs of specific groups of users before it has fulfilled its primary mission.

Following on the dangers of INTELSAT losing sight of its primary mission is the question of the growing promotion and creation of alternatives to the INTELSAT facilities. As the director general of INTELSAT pointed out in testimony before the United States Senate, much of the development of alternatives results from its inability to provide public services in various parts of the world. These existing alternatives, in Palapa, in Arabsat, in Eutelsat, have generated new proposals for alternatives to INTELSAT which seek to fill its public telecommunications role in regions vital to INTELSAT's continued operation, such as the Japanese proposal for the Pacific. Whether such direct competitive proposals are in the best interests of all users and the INTELSAT system is a question that may need to be examined and debated both within and outside of INTELSAT.

At the moment, however, most nations do not appear to be amenable to direct competition with INTELSAT in its primary role. What some propose is new arrangements to meet new needs, both in services and facilities. As officials of the European Space Agency have observed, the INTELSAT system probably has reached the point where no additional economies of scale exist; new and separate facilities under different management can meet specialized "utilizations" more efficiently and effectively.

NEED FOR FLEXIBILITY AND CHANGE

Given the increasing reliance of international commerce on communications, inefficiencies and inadequacies in communications facilities will present operational problems and opportunity costs which can slow down or stifle economic growth and competitiveness. No nation can afford such damage to its economic development if it can be avoided. And, given the evidence of experience, there is little doubt that restricting the growth of alternatives will result in the public telecommunications service suppliers ignoring specialized needs. In the United States, only after the authorization of user-owned communications facilities as alternatives to the public network were a wide range of business needs and demands met. Moreover, as Bank of America noted in its recent letter concerning the Orion Satellite Corporation proposal to the FCC,

> Current development in United States telecommunications . . . demonstrate the limitations of existing international . . . satellite offerings. Satellite domestic satellite telecommunications can be far more effective and responsive to user needs than the services provided by the existing international telecommunications structure. One such alternative is embodied in the proposal by Orion Satellite Corporation. Orion would offer Bank of America

the opportunity to own group and space equipment for telecommunications to and from Europe. It will permit the bank to communicate directly with its offices and facilities through user-owned, on-premises facilities, relieving administrative burdens and providing unprecedented flexibility and reliability.

Stated simply, without the development of alternative facilities, the economic activity dependent on telecommunications and the economic growth it represents will suffer.

After all, telecommunications is a servant. That truth is often forgotten when people begin debating about how communications facilities should be structured, who should operate them, or how they should be used. A recent report on International Telecommunications and Information published by the Foreign Relations Committee of the U.S. Senate highlights the fact that telecommunications is not an end in itself, that it exists solely to support other economic and social activity. In the words of the report, "data processing, telecommunications, and other information technologies provide the underpinning for increased productivity and growth in other industries and for continuing overall economic development."

International telecommunications provides an essential support system for the commerce that fuels the economic growth of the entire world. Without efficient communications at reasonable cost, international finance, multinational manufacturing, and a variety of increasingly important forms of counter trade could not exist. Moreover, without increasingly efficient and effective telecommunications systems, ever more sensitive and adaptable to the needs of end users, the continuing expansion of international commerce vital to national growth and development cannot be sustained. For telecommunications to remain a good servant, it must adapt and change as it grows.

CONCLUSION

The continuing revolution in satellite communications technology can provide even greater gains in the next twenty years than were provided in the last twenty. But to fulfill even half its promise, the technology cannot be fettered by institutional arrangements grown rigid. In the last few years, INTELSAT and those who manage it through their dominant investments seem to have begun to develop some of that potentially crippling rigidity. Particularly when diverse designs to meet diverse needs have become both possible and economically attractive, it would be unfortunate if INTELSAT were suddenly to assert a monolithic role that can only hinder development.

The governments who have given INTELSAT its license to grow and thrive should take steps to assess the consortium's capacity and commitment to meeting its primary mission of providing international public telecommunications services on a global basis. In particular, they should examine what role, if any, INTELSAT should play in supporting services beyond those intended to provide international connection for the public switched network. Should INTELSAT divert scarce resources to meet specialized needs not of direct benefit to its primary mission? Far more important, should INTELSAT be permitted to campaign against new system proposals where those systems would not intrude on INTELSAT's primary mission of providing facilities for message telephone and related switched services?

S I X

The Entry of New Satellite Carriers in International Telecommunications: Some Interests of Developing Nations

DOUGLAS GOLDSCHMIDT

The possible entry of new communications satellite carriers into international markets has led to questions about the benefits and costs of these new carriers to developing countries. These questions largely arise from INTELSAT's claims that it will suffer significant financial harm from entry, and from the counterclaims by the applicants that their services will both benefit new customers and not harm INTELSAT. Only one of the applicants, Panamsat, has designed its service primarily for service to developing nations. Thus, one must ask whether the other entrants, which will primarily serve the United States and Western Europe will have any affect on, and anything to offer, the developing world.

These issues are important to developing nations. There is no question that developing nations have a strong interest in an international satellite system which provides the types of connectivity INTELSAT now provides. While one might argue with facets of the INTELSAT system, it clearly is popular with most of the developing world.

There is also no question that many of these nations are interested in developing domestic and regional satellite systems for telecommunications, broadcast networking, and, increasingly, direct broadcast services. Domestic satellite systems exist, or will soon exist, in Indonesia, India, Brazil, and Mexico, and studies for systems are going on in Pakistan, Turkey, China, Argentina, and the Andean group, to name a few. Apart from Indonesia and India, all other domestic satellite systems in developing nations are provided using leases on INTELSAT satellites.[1]

The policy interests in this area are in the roles that might be played by INTELSAT, regional satellite carriers, and private satellite carriers in meeting the developing nations' service demands. There is also the question of whether the entry of private satellite carriers, operating either in new markets or in competition with INTELSAT, will adversely affect INTELSAT's ability to provide service to developing nations. The best point of departure is to review the issues tied to international telecommunications linkages, and then move to the problems of regional and domestic satellite service.

INTERNATIONAL TELECOMMUNICATIONS LINKAGES

The INTELSAT system was designed to make maximum use of global connectivity for telephone and other switched services. By using a small number of earth stations within each country (generally one per primary path satellite) operating through a small number of satellites, it is possible to provide maximum connectivity among all member nations by increasing the number of individual links or routes served by each satellite.

The INTELSAT system provides wide connectivity at a price. Achieving wide connectivity among areas that do not have strong mutual communities of interest requires wide beam dispersion. As the beam is dispersed, the EIRP diminishes, requiring larger and more costly ground stations, or less efficient use of the space segment. In addition, the INTELSAT satellites are based on standard designs oriented to meeting the requirements of the Atlantic Ocean region. These requirements are different from the other two regions and from the non-international services. While INTELSAT has expressed interest in developing satellites designed for specific purposes and areas, it has not yet done so.

This engineering design has not been a serious problem in developing the global INTELSAT system. However, it does pose problems for regional and domestic services, which will be discussed below.

The INTELSAT system's success is demonstrated by the major changes and improvements in international telephone service to developing nations since INTELSAT's creation. In the past long-distance telephony to developing nations was often routed through third countries, generally in Europe. This meant that neighboring countries might have to call each other through London, using HF radio links of poor quality. Capacity into these countries was limited, costs high, and quality unreliable. While some cable systems have been available to developing nations, these have been relatively limited, and have been used to connect high density routes.

Developing nations now largely have reliable, high quality telecommunications links with their neighbors and major trading partners. Their international telecommunications services have tended to be the most profitable part of the overall telecommunications activities. And they have far greater control over their international communications than in the past.[2]

INTELSAT's political structure was a major factor in promoting its penetration into most countries. While the organization is dominated by the United States, there has been an attempt to emphasize the organization's nonpolitical nature and its structure as a cooperatively owned, nonprofit venture.[3] The cooperative structure tied with the separation between the space segment, owned by the cooperative, and the ground segment, owned by each national signatory, helped ease many of the political obstacles which have traditionally occurred in organizing other regional transmission systems. Countries can join the INTELSAT system without many of the political problems associated with arranging bilateral arrangements with neighboring or other states, or with systems explicitly tied to national or political interests, such as Intersputnik.

However, INTELSAT's success does not mean that it has satisfied all demands for service, among both the industrial and developing nations. Among the industrial states there has been growing interest in developing new international satellite systems for providing specialized communications, not served well by INTELSAT, and developing nations have been searching for more effective means of providing regional and domestic communications. It is the proposed entry of new carriers centered in the United States, but likely to be joined by European firms, which has made the debate of how international satellite communications should be provided, more pressing.

PROPOSED ENTRY INTO INTERNATIONAL SATELLITE MARKETS

Since Orion's application to enter the international satellite market was filed with the F.C.C. in 1983 four additional applicants have indicated interest in the market. Volumes of testimony, pro and con, have been filed in Congressional hearings. This paper will not dwell on the merits or substance of these applications—the papers by Mr. Vizas and Mr. Pelton discuss the issues posed by the new entrants to the satellite market.

The reason for the applications for new satellite service is the perception by some entrepreneurs that INTELSAT is not economically meeting the needs of particular users. The calculation of potential markets is based largely on INTELSAT's engineering economics. As the space segment has largely determined the engineering, the ground segments have been expensive relative to the ground segments usable with domestic satellite systems offering higher EIRPs. Lower ground segment costs can quickly overcome space segment costs for domestic satellite systems, or for systems providing specialized services, like Customer Premises Services (CPS), which require many inexpensive earth stations.

Also, INTELSAT's operation of its satellites for the Atlantic region leads to a spacecraft design that is not as economical for domestic or regional services as a satellite specifically designed for such services. By changing the engineering of the overall system specifically to meet certain types of services, significant economies may be achieved in overall costs.

Whether there is a demand for satellite services which is constrained by INTELSAT's costs and which would be stimulated by a lower cost package cannot be verified before the service is actually offered. However, the history of satellite services in the United States and INTELSAT's own traffic history seem to suggest that lower costs will lead to large increases in customer demand.

INTELSAT has raised a number of objections to the applications for new entry, largely focusing on the potential for economic harm to INTELSAT, with the potential of significantly increasing INTELSAT charges, and the possibility of collapse of the INTELSAT system.[4]

Economic harm can occur in two ways—through direct diversion of traffic which is a significant part of INTELSAT's switched telecommunications and video traffic, and through the diversion of secondary traffic (i.e., specialized traffic, private lines, and some video). While two of the five potential entrants might divert some of the primary traffic, INTELSAT's attention has been largely focused on the secondary traffic.

In a recent study prepared at INTELSAT's request by Hinchman Associates, the problem of economic harm from this "secondary" entry is directly tied to the structure of INTELSAT's major Atlantic Ocean services. That is, new generations of satellites must be placed in the primary path to accommodate the rapidly growing service demands, leaving INTELSAT with older, but still usable, satellites. These satellites are used to provide domestic and specialized services, and their revenue contributions help to diminish the overall INTELSAT revenue requirement. As Hinchman argues: "so long as INTELSAT has available residual or excess capacity sufficient to satisfy additional demand, the satisfaction of that demand by other systems will necessarily result in higher costs per unit of capacity actually provided and utilized by INTELSAT members—including capacity utilized for basic telephone and telegraph services."[5]

Acceptance of this argument requires first, that IN-TELSAT's assumption that any of the new carriers will divert substantial amounts of traffic from the INTELSAT system be correct. For example, Orion and Panamsat have argued that the traffic which they hope to carry is not presently carried by INTELSAT, and will not be carried by INTELSAT in the future. Orion argues that the economies of the ground segment possible through specialized satellites cannot be realized using INTELSAT's spacecraft and, given the total costs of using INTELSAT for CPS, the service will not develop.[6]

Similarly, Panamsat has argued that regional broadcasting in South America has been constrained by the high costs of the INTELSAT occasional video leases and the cost of the ground segment. Panamsat anticipates that the substantially lower costs associated with its planned satellite, with high power coverage of specific South American markets, will stimulate the creation of a large market for regional and domestic broadcast networking.[7] While INTELSAT has noted on several occasions the importance of reducing ground segment costs associated with its system, it has not addressed this aspect of demand elasticity in its public testimony.

If, in fact, INTELSAT is not carrying the traffic proposed to be carried by the new entrants, but would like to carry the traffic now that it has been identified, it is difficult to argue that serious, or any, economic harm is probable. This would hold even for the extraordinarily narrow definition of harm proposed in the Hinchman study. The difficulty with the argument is exacerbated if static market definitions are rejected and the possibilities of overall market expansion taken into account.[8] The definition of economic harm which the Hinchman study proposes has no relevance in a competitive market and, interestingly, was rejected by the Federal Communications Commission when raised by AT&T during the early 1970s specifically because it posed an overly narrow and static view of the telecommunications marketplace.

Second, one would have to argue that the growth in INTELSAT's primary services will not generate sufficient traffic to meet the system's revenue requirements. However, INTELSAT's own predictions of traffic growth in its primary services—switched telephone, data, and telex, as well as television transmission—are all projected to show substantial gains over the rest of the decade. It is far more likely that any losses from these figures will be from competition from new cable capacity in the Atlantic and Pacific, which will directly challenge some current INTELSAT markets, than from new carriers attempting to serve markets which either do not currently exist or which are marginal.

Third, one would have to assume that INTELSAT has little control over its current plant or future investment. This is a difficult argument to make with satellites which are medium term investments. Given the ten-year life of satellites, INTELSAT has far more flexibility to modify its investment decisions than a firm like AT&T, which has far more embedded plant with economic lives of greater than twenty years. INTELSAT can move satellites, control launching dates, cancel satellite production, and modify satellites, all admittedly, at some cost. Again, this is far more flexibility than AT&T had at the time competition was allowed by the F.C.C. during the 1970s.

Fourth, one would also have to assume that INTELSAT has virtually no control over its business practices or pricing. While INTELSAT's structure makes changes in business practices difficult, it does not make them impossible. In the face of the types of economic harm INTELSAT has alleged will occur due to new entry, it is likely that its owners will agree to changes in practices, particularly pricing policies, to maintain the firm's economic health.

It is useful, however, to review the possibility of economic harm to developing nations if INTELSAT's assumptions about traffic diversion are accepted. The Hinchman study commissioned by INTELSAT projects that INTELSAT's transponder costs will increase between 8.6 percent and 35.6 percent by 1987, depending on the level of traffic diversion due to new entry.[9] Transmission costs represent less than 10 percent of the end to end costs of an international communication.[10] The vast majority of the costs are embedded in the ground segment. Thus, even with the worst case increase in the space segment costs, the actual increase in overall costs for international telecommunications would be approximately 5 percent. It is also worth mentioning that in many countries, particularly developing nations, the international telephone service is used to generate large cash surpluses. It is well known that the rates for overseas telephony are far in excess of costs, and that considerable profits are generated by the overseas service.

One major reason for this is that demand for international communications is relatively inelastic in countries with high telephone prices—the primary users are likely to be businesses for which communication is a minor part of the production function. Thus, raising the prices further will not significantly affect demand. Reducing prices dramatically will stimulate demand, but this raises other problems outside of the interests of the immediate debate. It is unlikely that a 5 percent increase in costs would be noticed within the prevailing overseas rate structures and could probably be passed on to its users.

The far more troubling intimations that the overall INTELSAT structure is threatened[11] are improbable viewed in light of INTELSAT's own traffic projections for its primary services. A serious threat to INTELSAT's structure would have to be predicated on massive diversions of traffic, not just from the North Atlantic route, inability to change investment programs or plant utilization, inability to change pricing for services, and a specific decision by major signatories to allow the organization's collapse. This combination of factors does not presently exist, and is unlikely to exist in the foreseeable future.

Given the current data available both from INTELSAT and its competitors, it is unlikely that the entry of new satellite carriers will adversely affect INTELSAT's service to developing nations in any significant way in the forseeable future. Even in the worst case posited by INTELSAT's economists, the cost increase for total end-to-end international communications will be modest. However, other developments which may pose serious problems for the INTELSAT organization, unless it begins to modify its approaches to engineering and organization, may arise within the decade by the growth of new international cable systems, and the probability of new regional satellites offering transborder switched communications.

REGIONAL AND DOMESTIC SYSTEMS:
THE SEARCH FOR CONNECTIVITY

Like the industrial nations, developing nations have been seeking means of promoting communications outside of the INTELSAT system. Both regional and domestic satellite systems have been the focus of intense planning. Systems exist in Indonesia and India, and will soon be implemented in the Middle East, Brazil, and Mexico. Other systems are being examined for China, Africa, Pakistan, Thailand, the Andean Region of South America, and Argentina, to name a few.

The development of national and regional networks can be problematic in the early stages because of the need to develop both the ground and the space segments within a short time period. Satellite economics are "lumpy"—the initial investment is relatively high, inflexible, and in space. Its efficiencies, in effect, are achieved in "lumps" of utilization—the greater the utilization, the greater the efficiency. The failure to rapidly develop the ground segment, by far the most expensive part of a satellite communications system, or to develop a sufficient customer base, means that the space segment will be poorly utilized and costly.

This was a familiar problem in the development of the U.S., Canadian, and Indonesian domestic systems. In the initial years, and during subsequent periods of excess capacity, space segment was utilized inefficiently, with attendant financial penalties to the carriers. This was particularly a problem for the U.S. domestic satellite carriers prior to the emergence of the television distribution market.

One solution to this problem has been to utilize spare satellite capacity, when available, to develop domestic services. This has allowed nations to lease the amount of capacity required in the short term, allowing the gradual construction of ground segment and development of customer bases. INTELSAT has been able to provide domestic service due both to its requirement to maintain backup satellites in the event of the failure of one of its primary satellites, and its possession of functional satellites which have been replaced by new, larger capacity satellites.[12] These older satellites often have a number of years of economic life remaining and can be profitably leased by INTELSAT at a price higher than their short-run marginal cost. More recently, the Indonesians have made domestic leases available for its ASEAN partners on its spare capacity on the Palapa A-1, and will presumably make capacity available on the B-1 once the B-2 is placed in service.

THE INTEREST IN DOMESTIC
AND REGIONAL SERVICES

Domestic and regional satellite systems are developing outside of INTELSAT for a number of technical, political, and economic reasons. Nontechnical reasons for domestic and regional systems are national or regional pride, attempts to promote national or regional cohesion, development of regional or domestic high technology industries, and an interest in having greater control over national or regional communications than is possible through an international organization. The validity or importance of these reasons is not of concern here—what matters is that a growing number of nations are using these arguments to justify system construction.

On the technical level, INTELSAT's limitations make domestic satellites attractive investments, assuming the need for a nation to increase the capacity and scope of its telecommunications system. INTELSAT's domestic leases are clearly a secondary use of its satellites. These satellites were largely designed to meet the requirements of the Atlantic Ocean region. As a result, multiple access to the satellites using inexpensive ground segments is not possible except, at least at the moment, with the tolerance of substantial inefficiency in the use of the space segment. Other types of satellite communications, for example, use of small, inexpensive television receive-only earth stations, are problematic due to the low EIRP. There are also relatively high ground segment costs caused by the requirement for circular polarization with the new INTELSAT satellites. Thus, large satellite-based domestic systems are to a large extent precluded by the cost of the ground segment. Also, INTELSAT's use of retired satellites for domestic service has been problematic at times because of operational problems attached to the aging satellite.

For traffic among neighboring nations where multiple gateways may be desired, for extensive telecommunications development, as well as for extensive broadcasting use, the higher power available from a regional or domestic satellite with focused or high powered beams offers large savings in the ground segment. Ground station intensive systems like those in Alaska and Indonesia are economically dependent on the high EIRP possible from focused beams. Extensive rural systems, which must have very low costs to be economically viable, would be very difficult to build using INTELSAT's capacity.

INTELSAT has attempted to improve its service offerings for domestic service, and particularly for rural services, over the past year. For example, one major new tariff offering, the Vista service, was designed to meet the perceived requirement of developing nations for rural services requiring less than one quarter transponder leases (the minimum capacity one can lease). The Vista service offers relatively low space segment charges on a channel basis, and allows the use of five meter earth stations, inexpensive by INTELSAT standards. This service however, remains expensive in comparison to what is achievable with domestic or regional satellites largely due to the cost of the ground segment. Recent estimates for earth stations for use in Australia were in excess of $200,000 for two SCPC channels. A comparable station produced in small quantities for a domestic satellite would be less than $100,000. Unless significant cost breakthroughs are achieved in earth station technology, the least expensive ground segment and hence most efficient domestic satellite systems will be those using satellites with greater beam power than INTELSAT now offers.

INTELSAT could improve its domestic and regional services through changes in its engineering. It has considered providing higher powered beams or even satellites tailored for domestic purposes without having reached any decision. Such a decision could represent a divergence from its primary responsibility to provide universal telecommunications service and would increase the organization's levels of risk. At the same time, it might be able to provide domestic and regional services more quickly than a new regional or even domestic satellite organization. Whether such new approaches to its business would be acceptable to IN-TELSAT's members needs to be explored.

MECHANISMS FOR CREATING DOMESTIC AND REGIONAL SYSTEMS

While there may be compelling reasons for developing national or regional systems, often economics and politics interfere. It is in these areas where the possibility of new entrants may be useful for developing nations.

For domestic systems, the most serious economic impediment is lack of sufficient traffic to justify a dedicated satellite system. While one may project growing into a satellite, the expansion of the ground segment may be sufficiently expensive to make a satellite cost ineffective for some time. This, as discussed previously, is one reason for the success of INTELSAT's domestic leasing program. Within the past two years Colombia, which had proceeded quite far in developing a domestic satellite system, decided to withdraw largely due to the cost of developing a national system. The two existing systems in developing nations—India and Indonesia, exist as much for national political purposes as for any economic motivations. Thus, a regional or global system (i.e., INTELSAT) can provide a country with interim service until it has sufficient traffic for its own system.

A dedicated regional system can provide higher powered beams than INTELSAT, making it more attractive for intensive development of the telecommunications system. Regional systems offer the greater likelihood of being able to develop sufficient traffic to economically justify the satellite. However, creating regional systems introduces political and administrative issues which can greatly extend the amount of time necessary to organize and implement the system. Such issues include resolving who will invest how much, who will receive which components of the system's operations, and so forth. As the Arabsat case showed, and as now can be viewed in the intense politics surrounding the proposed African regional satellite, this is not a trivial set of problems.

Four vehicles for regional service may usefully be explored. First is the approach now being explored by the Andean nations through the ACETA mechanism to organize regional service by collectively leasing a group of transponders from INTELSAT. This mechanism will minimize the initial investment for the participants to the ground segments and only the necessary space segment. It will also allow the development of a technical organization which will eventually be able to take over the major technical operations of a satellite system.

The Andean countries are hoping to be able to completely take over the leased transponder's use so that they have fairly complete control over the service. As long as they do not interfere with other INTELSAT services, this will allow the countries to utilize nonstandard technologies, with presumed savings in ground segments. The long-term goal of this approach, however, is to develop a regional satellite apart from INTELSAT.[13]

The second mechanism is the creation of a regional satellite organization, like Arabsat or Eutelsat, which owns and operates its own system of satellites on behalf of regional members. These satellites are designed to make optimum use of beam power within the region. While this mechanism is appealing on paper, the long lead time required for Arabsat, the current political disagreements within Africa, and the mixed experience in South America with regional collaboration may make this approach a long-term, rather than medium-term, means of developing satellite services.

The third mechanism is the expansion of a domestic system into regional service. This has occurred in a limited way in the Western Hemisphere, with the U.S. domestic satellites being used for limited transmission of video services to the Caribbean, and occasional use of the Canadian satellites by the United States. On a larger scale, Indonesia has applied to INTELSAT for permission to use the Palapa for regional service within the ASEAN group, now uses it for transborder service, and has successfully leased service to its ASEAN partners. The Palapa B series is capable of providing domestic services to all of ASEAN, as well as to New Guinea.

The limitations on this type of service relate to problems of regional politics. As Palapa is owned and operated by Indonesia, this raises some questions in the other ASEAN countries about the long-term risk of trusting their domestic communications to a country with which they may eventually have political differences. While ASEAN is a relatively cohesive group now, this is clearly a concern in at least one of the states.

A fourth mechanism is through the development of a private regional system, such as Panamsat is proposing. In this scheme, individual countries, or groups of countries, can arrange to buy or lease capacity from Panamsat. As with the Andean model, they can acquire the capacity needed in the short term, while developing regional or domestic mechanisms for an eventual satellite. A major difference between the two is that the Panamsat satellites will be designed to provide significantly higher powered coverage of South America than the capacity the Andean countries may lease from INTELSAT. Panamsat has argued that the overall design of its system will allow lower space segment costs and, more importantly, much lower ground segment costs than would be possible through use of INTELSAT. Also, the countries participating in the Panamsat program will be able to own their transponders (which can have important tax and regulatory benefits), presumably providing greater control over transponder use, and will be better able to utilize customized services, given the greater EIRP.

While Panamsat raises the political problem of being owned by entrepreneurs who are seeking profits, of being separated from the traditional means of providing international telecommunications service, and having a far higher level of risk for the participants than INTELSAT offers, it does offer the advantage of coming into existence relatively quickly with satellites tailored for very specific types of service to South America. Its service offerings are sufficiently open that each participant would have control over its space segment. It also presents no long-term impediments to regional or domestic system developments.

It is possible that offerings like Panamsat will be evolutionary businesses, changing as regional and domestic traffic, politics, and economics allow the development of new satellite systems. While this may pose some long-term problems for Panamsat's investors, this is not a problem for the initial policy analysis. Unlike INTELSAT, Panamsat would be constructed strictly with private funds, so that any loss of capital due to new systems would encumber only the investors, not other users of the INTELSAT system.

As matters of policy, private systems such as Panamsat should not raise different problems relating to economic harm than such regional systems as Arabsat or Eutelsat. The success of such systems will necessarily rest with the judgment of the countries in the region that such service is preferable to the other alternatives.

CONCLUSIONS

New entry into international satellite markets is unlikely to jeopardize the interests of developing nations in achieving global connectivity via INTELSAT, or to cause significant increases in end-to-end costs. While the proposed entrants pose significant deviations from the means satellite service has been provided historically, these deviations will have to be accepted by individual nations if these ventures are to succeed. The success or failure of the ventures will rest with private investors.

More importantly, at least one of the new entrants is proposing service solutions for developing nations which should prove more cost effective for domestic or regional services than INTELSAT's offerings. The diversity of services and different cost structures which may come through new entry may help expand the international, regional, and domestic satellite markets in developing nations in ways similar to what has occurred in the United States and Canada, and more importantly, Indonesia. Given the enormous requirements to develop an information infrastructure in the developing world, the possibility of new service offerings may have enormous development potential.

There is no reason to suggest that INTELSAT be excluded from these markets, or that it not attempt to tailor its offerings specifically for these markets, assuming that any new service offerings would not affect its primary international services. However, INTELSAT neither offers appropriate capacity to provide many of these services nor has it indicated when it will. Given the enormous technological changes in telecommunications which have made many services affordable given appropriate engineering and service configurations, individual nations, or groups of nations, should be offered the option of having diverse vendors of satellite services for their regional or domestic needs. Private vendors offer one approach to the attempt to meet these needs.

NOTES

1. Developing countries using INTELSAT domestic leases include Algeria, Argentina, Brazil, Chile, Colombia, India, Mexico, Niger, Nigeria, Peru, Sudan, and Zaire.

2. At least to the gateway station. Many developing nations continue to have problems with their terrestrial telephone systems, so that calls may have difficulty getting from the international gateway station to their final destination in the country.

3. While INTELSAT is technically a nonprofit cooperative, it still earns its cost of capital, similar to regulated utilities. Given that the members of the cooperative are all state owned or private, publically regulated utilities, it is likely that INTELSAT's economic motivations are closer to regulated telecommunications utilities than to nonprofit charitable organizations.

4. See Director General, "Report to the Assembly of Parties on New Developments Concerning International Satellite Communications," INTELSAT Document AT-8-9E W/10/83, July 29, 1983.

5. Walter Hinchman Associates, Inc., "The Economics of International Satellite Communications," prepared for INTELSAT, May 1984, 1:13.

6. Christopher Vizas, "Letter to All Parties to the INTELSAT Agreement or Their Representatives at the Assembly of Parties Meeting," September 30, 1983.

7. Interview with Frederick Landman, president of Panamsat, November 18, 1984.

8. The issue of economic harm is highly problematic with INTELSAT as it wishes, as the monopoly supplier, to assume for itself the right to determine what is and is not harmful. This is awkward from any regulatory standpoint. It also appears to lead to inconsistent policy criteria, assuming INTELSAT's definition of harm, where traffic may be "lost" from domestic leases as domestic satellites are launched, and from regional carriage as regional satellites are launched, yet not be defined as economic harm, but be defined as harm if private carriers attempt to enter similar markets.

9. See Hinchman, "The Economics of International Satellite Communications," p.14.

10. Statement of Richard R. Colino, Director General-Designate, INTELSAT, before the Subcommittee on Arms Control, Oceans, International Operations and Environment, Senate Foreign Relations Committee, October 19, 1983.

PART III: NATIONAL PROGRAMS AND PERSPECTIVES

S E V E N

The Theology of Satellite Television: Dogmas That Are Holding Up the Progress of Satellite Television

BRENDA MADDOX

A long time has passed since Oscar Wilde or Bernard Shaw or Winston Churchill, it doesn't matter which, said that Britain and America were two nations separated by a common language. It is a paradox which is growing truer. The languages are, according to Richard Burchfield, editor of the Oxford English Dictionary, growing further apart. His reasons were, I think, the increasing heterogeneity of America, especially its acceptance of the Spanish language, and also its fondness for technical and psychological terms.

This persistent disparity should reassure those who like Mr. Jack Lang, the French cultural minister, fear the loss of national cultural identity under the great tidal wave of American entertainment, now borne farther than ever by video and satellite as well as plain old movies and television.

But it tends to be forgotten by those of us who think of ourselves as inhabiting one English-speaking television world.

We conveniently ignore the division, the sacred ideas, the facts of politics and geography that divide us and make developments on one side of the Atlantic very alien to the other, in spite of the satellites and cables that bind us.

This misunderstanding is a luxury that can no longer be indulged. Both sides of the Atlantic want to sell things to each other—home earth stations, consultancy services, television programs. And Europeans want, just as Americans watch the progress of the weather from Pacific to Atlantic coast, to see what is brewing up in the West to hit them next.

So what are the blind spots? What does each side fail to see when it looks at the other? I'd like to offer my personal view, based on the only thing journalists are expert at: asking questions.

Here is what I've gleaned. For a start, neither side appreciates the other's geography. Europeans do not even know the *song* that says "from sea to shining sea." They do not know how big the United States is; they do not feel it in their bones. This means that they do not understand at all the concept of broadcasting based on cities, which leaves large areas in between to catch-as-catch-can by cable television or home earth terminal.

Americans, from where they sit on the map, believe what they read in the papers. They see Europe as a "common market"—one big salesplace, not as big as the United States but not so very much smaller and very much alike in prosperity and tastes, for do not they all, Swedes and Britains and Italians, love *Dallas* and *Dynasty*? Americans forget about the plurality of governments in Europe. They would not believe (and it is hard to find out) the contradictory array of laws and restrictions on broadcasting and advertising: Italy bans pet food; Britain, almost anything below the belt, and Belgium, advertising itself. And they do not appreciate the protectionism and non-tariff barriers that go into preventing all kinds of things being sold across the borders of countries committed to free trade, at least with each other.

These blind spots converge when it comes to satellites because these involve television itself—that sensitive subject which, more than motherhood, is regulated by national laws which are derived from the ideas which these very different societies hold most sacred.

And these sacred ideas are in conflict between the two sides of the Atlantic. Americans place primary importance on freedom of speech and have placed it among the amendments to the Constitution, and they are proud of deregulation, Europeans and British, especially, believe in national sovereignty over broadcasting and its corollary—public service broadcasting; that is, the use of public funds to supply the entire population with a range of programs according to some ill-defined but powerful sense of what is the national good. And they are proud of planning for social change. Each side interprets the other's sacred cows in the crassest, not to say most cynical, light.

Let me start with *deregulation*. In Britain, this American phenomenon means letting market forces rip. Americans are doing it because they respect money-making, and only that, and want to get government out of the way. (This is not entirely false, like most prejudices.) It is seen by broadcasters as a cynical abandonment of the interests of the impoverished minority viewers. When Mr. Mark Fowler says, "the public interest is the public's interest," he is seen as disdaining all those television viewers who cannot vote with their pocketbook—the old, the young and the disabled.

That there is no awareness that there are other forces behind deregulation comes—as I see it—from an inability of social planners to foresee change. Thus they do not see deregulators' wish to smooth the introduction of new technologies or the disenchantment with government's past policies that may have brought about unintended or undesirable results. Deregulation is thought to have little to offer Social Democrats. It is not appreciated that one of the important forces behind American deregulation was when the Federal Communications Commission, under a Democratic administration, realized that all it had achieved in a decade of labored rulemaking was to protect the broadcasting industry from economic harm for which there was no evidence.

As a result of such misunderstanding we get deregulation European style. An example of the misinterpretation of deregulation is the Conservative government's privatization of British Telecom. They have simply turned the most powerful monopoly in the country over to the private sector, with puny regulation and no competition at all, except one hand-picked and deliberately enbelled competitor, Mercury. That is giving the communications revolution over to those who can make money out of it, the consumer be damned. Most likely, British telephone service—those call boxes you wrestle with at Heathrow—is not going to improve. Yet they think they are following America's lead in breaking up AT&T.

But fair's fair. The Americans look at public service broadcasting and see paternalistic, elitist control and condescension; a few decide what the masses should be believing. A lot of truth in that, too. As my uncle in Middleboro, Massachusetts, once said to me, pityingly, "The BBC? That's all educational, isn't it?"

It is *not* and it is not just the BBC. Commercial television in Britain now considers itself public service broadcasting and, while it may seem self-serving, it wants protection, like the BBC, against newer forms of television entertainment—it certainly submits to heavy regulation, even censorship at times; it hands over two-thirds of its profits before tax to the government, and does all kinds of programs against its inclination, at prime time.

There are three things Americans ignore, even when they admire the strengths of *Jewel in the Crown* or *Brideshead* or *Monty Python*. One is that the public service obligation includes the duty to blanket the country—to the outermost Hebrides—with signals of good quality and, in the case of the BBC, with four radio services as well. There *are* no pockets of have-nots. All pay the same, the philosophy goes in the case of the BBC's license fee, and all receive the same.

The second is that in the small of Europe, people count on their national service as a unifying experience, one of the distinctive lines between themselves and their all too close alien neighbors. On Sunday nights there is a BBC program called *Did You See*, a discussion of television during the preceding week (not only the BBC's). The implication is that many people have been watching the same thing. They are certainly open to invitation to watch more channels than four, but they are comfortable with the easy-to-read television schedules which can be found in any daily newspaper and are good in any part of the nation.

The third thing Americans forget is that Europeans pay license fees—an annual tax on television households—they do *not* feel they have gotten their television for nothing. So, when faced, say, with pay TV, they feel that they are being asked to pay again for what they have paid for already.

These patterns of national broadcasting work against Open Skies policies. An interesting thing is that these philosophies are coming into collision because of satellites. If Home Box Office caught on like wildfire when it was spread across the United States to cable systems by satellite, why should it not travel 3,000 miles in the other direction?

European entrepreneurs quite as much as Americans made this speculation. France, West Germany, and Britain made plans for rapid cabling, to be paid for—directly or indirectly—by people's appetite for more video entertainment. But the governments forgot their own unwillingness to loosen the regulatory hold on the new choices that could be offered (and in West Germany, the right of each state to make its own rules on broadcasting, so that, until now, national transmission by satellite to cable systems, is forbidden). The explanation—in Britain at least—is that the two reasons cable grew in America in the first place do not exist: poor reception and a wish to see movies uncut and uninterrupted by commercials. The experience is not—strictly speaking—transferable.

Turning now to direct broadcasting from satellites, the difference in view is even more striking and more set for a head-on clash. Back in 1977 at the World Administrative Radio Conference, the Americans and Canadians recognized that they did not see DBS developing as other countries viewed it. They refused at that point to agree to the plan which gave most countries of the world enough frequencies for five channels of satellite-to-home television, and a place in orbit from which to beam them down.

Six years later, when they came to do their own regional plan, they congratulated themselves on their wisdom. Advances in receiver technology meant that you could reach small home dishes with satellites of much lower power than was dreamed of in 1977's philosophy. So these countries advanced on Europe to sell some of their satellites and they are mystified by the insistence of France, Britain, and West Germany to go ahead with programs of high-powered satellites with a range far wider than their national territories—all they are supposedly interested in.

I had lunch not long ago with an American aerospaceman who was shaking his head. "they could reach the same national audience with a medium-powered satellite," he said. "And they could get more channels."

"They don't want more channels," I said.

"They—don't—want—more channels?" he repeated after me, as if I had said they did not want any more sunshine than the meager ration they get already.

No, I had to tell him. They don't. They do not want more than three DBS channels to dilute the national television mix. But, and I speak about Britain, it is not just loss of sovereignty that they are worrying about. They argue that too many channels reduce choice—that more means less, that the only way to fill a dozen channels would be with filler material. And, they, believing this is an economic argument at heart, are sure that if viewers are to be wooed to buy or rent DBS dishes they must have an alluring alternative to what they already get over the air, and that means a DBS service of high quality. They just have not figured out what it is to be.

Who is right? We shall see. I think Luxembourg's Coronet project is the most interesting development in communications today. It is where West meets East—the raw force of the new world, if Tom Whitehead will forgive me—crashing up against the artifice of Europe. A medium-powered satellite with sixteen transponders to be rented out to whoever thinks they can make money on them (but no porn). The forces against it are formidable—the European PTTs plus the broadcasting monopolies, a double whammy—but so are market forces that are pushing it. The kind of high-powered DBS that the European Big Three want is proving too expensive even for their own narrow aims.

The advent of transborder satellite services brings with it another idea whose time may have come for Europe and something that is important for the medium-powered satellite. It is freedom of information. Computers plus satellites have created the possibility of instant access to information. Can European ingenuity find a way to stop it? The satellites are creating programs that waft over national boundaries. Can European laws stop commercials from crossing?

There are three developments that give grounds for hope that the old rules controlling information will be changed. One is that the EEC, the Common Market, wants to unify advertising standards so that cable and satellite programs can move freely across borders. The other is that there is a move among liberal lawyers in Europe to interpret Article 10 of the European Convention of Human Rights as a European version of the American First Amendment. It commits signatories to permitting freedom of expression with very narrow exceptions. Dangerous American ideas—such as that even a ban on liquor or cigarette advertising may violate the right to free speech, if you interpret that to include commercial speech—are now beginning to be looked at.

So Europe may be forced, by the democratic nature of the new technologies to loosen its stranglehold on who may know—or view—what. A New World Order triumphant!

Will the public's right to non-marketplace television—a decently financed public broadcasting service—cross the Atlantic in reverse direction? Of course not.

If I had to choose between the two, I'd choose freedom of information, freedom of government control of the content of television. But with a heavy heart. I'd be giving up much that has enriched my life and my children's.

Maybe that is too pessimistic. Someone may find a way to have deregulated television and quality too. Maybe, however, we have to accept that some ideals are incompatible.

E I G H T

The Doldrums of Europe's TV Landscape: Coronet as Catalyst

MARIO HIRSCH

The careful observer of developments in the fields of television and communication in Europe will be struck these days by what seems to be an amazing contradiction. On the one hand, one senses everywhere in Europe the awareness that something has to be done to react to the threat associated with what is widely seen as an hegemonial plot by U.S. interests in the field to take control of European developments. On the other hand, Europe is quite definitely caught up in political and industrial policy contradictions that prevent the continent as a whole from making use of its combined resources and creativity.

The Coronet project, initiated by the Luxembourg government under Prime Minister Pierre Werner in 1983, with the help of an American midwife, Dr. Clay T. Whitehead, highlighted in a unique way these contradictions. The fact that one of the smallest European countries had come up with what was unanimously acknowledged as a bright idea was not especially helpful since it was the source of quite some resentment. After all, you cannot expect, acting from a Luxembourg base, to get away unharmed, having proved to the larger European countries that they were heading in the wrong direction. Confronted with the Coronet concept, most of the other European satellite projects appeared as cumbersome, lame-duck undertakings where neither the technology nor the economics were quite what was required to meet the U.S. challenge and the needs of the market.

The Coronet project has run into considerable difficulties on the political and the regulatory front. Its promotors and the Luxembourg government underestimated the power of vested interests in the field of TV and communications. At the same time, Coronet has received support from precisely those sectors that are essential for the success of the project, namely, programmers and would-be programmers, electronics companies, advertising agencies, and a few of Europe's seedling venture capitalists. It remains to be seen how this bold project to expand television offerings in Western Europe will fare in the future.

THE EUROPEAN BROADCASTING PICTURE

Direct broadcasting by satellite has captured the imagination of European governments, satellite manufacturers, broadcasters, cable television operators, and film makers. The expectation is of course that millions of homes will receive extra television channels beamed from a satellite to a small dish aerial.

Most of the European projects, all of them government sponsored and tepidly supported by industry, have become aware of the fact that the risks are enormous and the start-up costs huge. These high costs, together with changes in technology, have called into question the suitability of the very high powered satellites envisaged originally for DBS. In Europe, the power levels for DBS were established at the WARC 1977 in Geneva. But since then there have been major advances in the technology, not so much in the satellites themselves but more so in increasingly sensitive and sophisticated reception equipment. The result is that it is no longer clear that Europe needs the high power satellites now on the drawing board and expected to become operational two years from now.

There are many indicators that Europe is having second thoughts about the DBS technology, this quite independently from the Coronet project. It is however fair to say that the emergence of this project has focused existing suspicions. It has prompted the conclusion that all these undertakings amount to what Brenda Maddox has termed a "desperate attempt in pursuit of the unviable" (in reference to the British DBS Unisat).[1] In France, the same point has been made quite conclusively in the famous Théry report early this year. This was an official investigation commissioned by the French Prime Minister into the TDF-1 DBS project which concluded that high power DBS was "passé." The European controversy is also fueled by the apparent contradiction between the ambitious cabling policy pursued by most of the major European countries and satellite broadcasting. It would seem that the cabling policy will make sense only if satellites are used to feed the programs they carry into the cable networks. However in order to achieve this, medium-powered satellites are all you need. This might explain the amazing success of the concept behind the European Communications Satellites (ECS) operated by Eutelsat. A much better bet would be, of course, the use of a medium-powered satellite of a kind still classifiable as DBS by the International Telecommunication Union rules. This is precisely the idea behind the Coronet concept. Such a satellite would allow both for individual reception and for the feeding of programs into cable networks as well as collective antenna systems.

It is a pity that Europe is not prepared, from a technological point of view, to give the right answer to these impending questions. It just so happens that Europe's aerospace industry has not yet caught up with the trend toward medium-powered satellites which seem to serve best its present needs. For that reason Coronet is obliged to rely on American hardware. But on the other hand, the lingering threat of Coronet has persuaded both the French and the German governments that it would be prudent to see the handwriting on the wall and they are thus accelerating the development of so-called "second-generation" satellites which will be most likely modeled on the Coronet technology. When one's detractors begin directly to imitate, it is a telling symbol of how Coronet has led the way.

Even more irritating to European governments, keen on keeping a tight control on broadcasting as they are, were the implications of technological advances for the regulatory environment. As a rule most European governments are not yet prepared to acknowledge that improvements in reception technology blur regulatory distinctions between high-powered DBS and medium-powered satellites, which operate under different rules but are still capable of delivering programs that can be picked up by individual homes. It is true that the medium-powered satellites operating in the Ku Band have not been designed for transmission to individuals. Also, they make use of frequencies not initially intended for direct broadcast to individuals, being considered point-to-point or Fixed Service facilities rather than facilities for general Broadcasting Service. But the improved reception technology together with the pressures of the marketplace have already made anachronistic the existing regulatory structure. Some of the problems Coronet has run into with European PTTs and Eutelsat are related to this lag, common in many technologically sophisticated industries, between actual practice and the regulatory environment.

Coronet, which prides itself on being the first private satellite television distribution company in Europe, has been seen, of course, as a major threat to state monopoly broadcasting prevalent in Europe. David Webster has proposed an excellent definition of the European way of doing things in broadcasting: "Europe has put its faith in a system which relied on the reallocation of resources in the name of the public good, by financing the bread from the revenues of a limited number of circuses."[2]

PUBLIC VERSUS PRIVATE BROADCASTING

There is an increasing awareness in Europe that the internationalization of broadcasting will not bypass the Continent. This means of course that existing structures will have to change. This applies especially to the state monopoly over broadcasting that has existed in most European countries. Satellite broadcasting techniques render national boundaries meaningless.

The Commission of the European Communities has urged EC governments in the recent "Green Paper on the Establishment of a Common Market in Broadcasting" to move toward a "common market for broadcasting" based on harmonized legislation capable of exploiting the looming expansion of radio and television transmission by satellite and cable.[3] The commission is of the opinion that the creation of a common market for broadcasting and cross-frontier distribution of broadcasting services will help push through the new information and communication techniques needed in terms of the economy as a whole. The possibility of a community-wide approach, including licensing to transmit via cable, the regulation of advertising, and the protection of minors has been widely hailed by most parties active in the field. The EC Green Paper recommends, for example, that advertising be limited to 20 percent of total cross-frontier broadcasting time. This is higher than several countries presently allow, but it is a measure of the advertising demand which the commission has identified.

The significance and prospects of this policy have been outlined explicitly in the document:

> Attractive broadcasting in the Community will pave the way for even more significant innovations in information and communication techniques. The cross-frontier distribution of broadcasting will provide listeners and viewers in the Community with new channels and programmes, which in turn are a necessary precondition for stimulating private demand to make use of the new transmission techniques. Investment of the order of over 100,000 million ECU in the Community as a whole will be required to establish viable information and communication networks. The main initial beneficiaries will be the whole telecommunications industry. The establishment of a viable infrastructure will create a need for new items of consumer electronics equipment, and private and commercial users of the information and communication infrastructures will require new and additional items of consumer electronics and office equipment. The demand for programmes will increase sharply, opening up new marketing possibilities for the originators of creative works and new employment perspectives for performing artists. Lastly, the commercial utilization of the new

communication networks will enable firms in the Community to increase their efficiency and cut their costs, as is essential if they are to maintain and improve their international competitiveness.

Despite this eloquent plea for an "open skies" policy in broadcasting, it must be added that for the time being this amounts largely to wishful thinking on behalf of the commission. True enough, some private groups like Coronet and Thorn-EMI (which directly or indirectly controls four out of twelve pay TV-channels in Europe), for instance, have decided to move ahead, even though the framework for their ambitious plans is still largely hostile. Market research undertaken by Coronet has identified a potentially promising marketplace that will develop from a total of 33 million European homes connected to cable today to 54 million in 1990 and approximately 70 million in 1995. But it is far from clear today whether things will proceed in this optimistic fashion or how open viewers (and their governments) will be to cross-border programming.

Major research done by the Economist Intelligence Unit and by CIT Research recently has raised further questions regarding the pan-European market and its growth prospects, pointing, among other things, to the uncertainty and unease with which the European cable industries currently view their development.[4] The suggestion here is that the entire European cable development program is in danger of being undermined by well-intentioned, but unrealistic governments, which are imposing technical, commercial, and financial burdens that the fledgling cable industry cannot support. This remark applies equally well to satellite policies. As it is, ends and means are proving everywhere hard to reconcile, as one notices a striking contradiction between the enthusiasm for high technology at almost any price and a lack of realism about subscription services, programming, and the investment picture. Unless the discrepancies between European policy and practice can be corrected, the impetus for these new developments may be lost. Although there is little pay TV yet in Europe, it is being looked on favorably by most countries eager to encourage cabling. At the same time commercial, trans-border satellite distribution is developing; and despite the fact that existing broadcasting is still reasonably cheap to viewers, the new techniques are challenging quite obviously the assumptions of European broadcasting and telecommunications policy and eroding the structure on which the existing industries and institutions have been built. The main factor in this erosion comes, of course, from the need to fund broadband cable development. Add to that the commercial pressure from program distribution and cable operators, the need for broadcasters to generate extra revenues, the redundancy of national regulations in the face of cross-national satellite distribution, the technical momentum in communications development, and one gets a good picture of the knots Europe must untie.

Most European governments are desperately trying to keep things under control. Faced with losing control over the television signals entering their countries, and thereby over the whole structure of their broadcasting (satellite signals are almost impossible to jam), most European governments have tried to set up broadcasting Maginot lines. Technical standards incompatible with those of other countries were and still are a convenient means to achieve this. As of today television standards remain incompatible in Europe and attempts to reach a common standard for DBS transmission have proved to be futile so far. The reason for this reluctance to embark on common standards can be seen in the fact that the present situation of incompatible standards suits some governments because it enables them to control the signals to be received in their countries.

Cable turns out to be an excellent national filter as well, insofar as it enables national governments to rig the market to inhibit the spread of satellite reception directly by individuals. As a matter of fact, it looks as if direct reception will be discouraged by technological standards and by an economic structure that makes cable cheaper for the consumer. Another reason why preference is given to reception via cable has to do with the fact that cable systems, being a public franchise, can be regulated ad libitum by government agencies.

European governments are slowly but inevitably recognizing that this leaves them with a troublesome problem. The recognition is dawning that without new, attractive, and suitable programming the ambitious largely government-financed cabling policies will most likely turn out to be a flop. In Europe programming, not channel capacity, is indeed the new scarce resource and most observers readily admit that the realities of commercial programming have yet to be effectively addressed. CIT research managing director Patrick Whitten has summed up the situation in a telling simile: "I believe that there has been too much concern about the quality of the road surface and not enough attention to where it is leading or the traffic it is carrying. We would like to see a thriving European cable industry with all its opportunities for new programmes and services and we believe it will develop. But our research suggests that it is being rather sidetracked."[5]

A good example of this sidetracking can be seen in the rather futile debate on American cultural hegemony. Coronet became a prominent victim of this lament, because it was widely but wrongly seen to pave the avenue for a forceful entry of American media companies into the European market. This is, of course, a caricature of Coronet's role. Coronet views itself as an instrument in the service of Europe's programming industry, which confined to the few public networks existing in most European countries has not had the outlets to grow. By offering up to sixteen transponders, Coronet will for the first time in Europe create unique opportunities for the whole production field. It is true, of course, that few European countries have the production base to feed multichanneled and transborder television for the simple reason that the production sector has been trimmed down to the limited needs of public broadcasting which more often than not is lacking the budgetary means to stimulate a genuine and lively home-grown production.

Because of this obvious deficit in programming, quite a few people are afraid that the only way to feed the insatiable and undiscriminating appetite for programs cable systems will be faced with, if they want to become attractive and viable, will be to rely even more heavily than now on U.S. programming resources and products. It is, of course, true that the only source of popular entertainment at the right quantities and at the right price is the United States. In 1983 the Commission of the European Communities published an interim report on "Realities and Tendencies in European Television"[6] which contained a shattering and embarrassing revelation about the foreseeable deficits in the programming industry in the advent of channel multiplication and a more liberal framework. On the assumption that on the average around thirty channels would be available in most European countries in the near future, the commission estimates the programming needs at 1.5 million hours per year. If you discard rediffusions and direct broadcasting of current events you are still left with some 250,000 hours of original programs. On the other hand, if you put together all the existing production resources in the television and movie business in the European Community as they exist today you barely manage to come up with programming not in excess of 2,500 hours per year. The gap is enormous and disquieting. In the "Green Book" the commission comes to a slightly more positive assessment of the situation that will prevail after 1990. Among other things it rejects the argument that the coexistence of two types of television organization—one financed from license fees and the other financed on a commercial basis—will inevitably lead to a drop in the high quality of programs.

THE CORONET CONCEPT

Jonathan Miller has argued persuasively that Europe can learn quite a bit from American experience:

> Europeans, who are still grappling to find an appropriate satellite strategy, would be wise to recognize the crucial role already played by satellites in energizing the American cable industry . . . By creating the possibility for economical distribution of new programmes to cable systems, and in particular, the distribution of pay-television services, the satellite carriers provided cable operators with an opportunity to double their revenue from existing subscribers, and an economic incentive to build new systems. It is no accident that the growth of cable in the past ten years tracks exactly the growth of the domestic satellite industry, and the new availability of satellite earth stations priced at a level that even the smallest cable operators could afford.[7]

Dr. Clay T. Whitehead, promotor, has also drawn the attention of Europeans to another development in the United States, a development he was himself instrumental in helping to bring about as one of the artisans of the Galaxy system developed by Hughes Communications. The new service emerging in the States and known as Satellite Master Antenna Television (SMATV) could turn out to be very promising in Europe if one keeps in mind the troubles the gigantic cabling policies have run into. SMATV is a hybrid between direct broadcasting and cable. The combination of medium-powered satellites and inexpensive earth stations opens the opportunity for small communities (they could be as small as a single apartment building) to establish their own self-contained program-distribution networks. The arrival of medium-powered satellites, which have an output of around 50 watts, is bound to create a new market for direct-to-home services. Coronet has not emphasized this dimension, largely to avoid provoking more European regulatory agencies, but everyone who is aware of the technical capabilities of medium-powered satellites knows that this possibility exists, provided the regulatory environment doesn't hamper it. Because medium-powered satellites have around twenty-four transponders, (although Coronet will only use sixteen to save the remainder as back-ups), they offer programmers an attractive device for clustering services, while at the same time providing consumers with an attractive supply of programming that will justify the purchase or rental of the necessary receiving equipment. Coronet is the only private enterprise satellite project currently underway in Europe. Unlike the government-sponsored satellite projects, Coronet is the only satellite television project designed technically, economically, and politically to tap the potential of the European commercial television market.

Coronet will provide the first private enterprise satellite for transmission of commercial television programs, including both pay TV and advertising-sponsored programs and new services such as teletext and direct delivery to VCRs. The system is designed to provide distribution to all three means of reaching TV sets: cable, collective antenna systems, and individual antennas. Coronet will have the capacity to provide between four to six audio channels for each video channel allowing simultaneous, multi-lingual service. Programmers will thus have not only their own natural linguistic market, but all of the western European market.

Unlike other European satellite systems that have been initiated for other purposes (aerospace subsidization, national broadcasting, or telecommunications), Coronet has been specifically designed to deliver commercial television to all European households in the most cost-efficient manner, either directly or through cable and collective antenna systems. This concept includes the satellite itself (which will be of well proven, reliable, commercial design and manufacture), the antennas (which will be small and affordable) and the common uplink facility in Luxembourg. Coronet is at the forefront in implementing the commercially viable technology in each of these areas. There may be further developments in antenna technology and manufacture which would lower the cost and allow Coronet to offer even more services than anticipated on current information.

Until recently, it was commonly believed that satellite transmission into antennas of less than one meter in diameter required very high-powered satellites. Conventional telecommunications satellites were thought to be useful only for sending signals into 4–5 meter antennas. This distinction has disappeared for all practical purposes as conventional satellites have become more powerful and as receiving technology has undergone significant improvements.

Specifically, because of recent advances in antenna and receiver design and better modulation techniques, medium-powered telecommunications satellites can transmit into antennas 0.9 meter in diameter with power levels of 50–53 dBW over most of Western Europe. This is only one-tenth the power of satellites considered necessary under WARC '77 rules.

The significance of this is that medium-powered telecommunications satellites can transmit television pictures of high quality into small antennas suitable for individual reception as well as into antennas at cable or collective antenna systems. These satellites cost less than half the price of higher powered ones and they have up to five times as many channels resulting in a dramatically lower cost per channel.

COMPETITION AND THE REGULATORY ENVIRONMENT

The Coronet technology will be optimized for TV transmission to all three means of video distribution: cable systems, collective antenna systems, and individual antennas. Coronet reinforces but does not compete with these three means of video distribution. Satellites have proved to be by far the most cost-effective means of sending video signals over a widespread geographic area to any or all of these ultimate consumer reception points. Most importantly, cable systems will need the large number of new programs brought by the Coronet satellite to become economically viable.

Laws and regulations concerning copyright and licensing of intellectual property are changing in Europe. The holders of rights to films and video material recognize that it is in their economic interest to secure a broad distribution base and they are working to ensure that the emerging rules in this respect are realistic. Programs are currently being transmitted by FSS satellites across national borders in both North America and Europe. This trend toward the acceptance of signals transmitted from across borders is likely to continue and to be encouraged by policy initiatives such as those outlined in the "Green Paper" of the Commission of the European Communities. Provisions in the Treaty of Rome concerning the free circulation of goods and services in Western Europe are cast in general terms and do not explicitly address questions of the right to receive television signals. The jurisprudence of the Court of Justice of the European Communities has however made it clear repeatedly that this kind of service falls under the provisions of articles 59 et seq. of the EEC Treaty which outline the principle of freedom to provide services.

Despite these provisos, the Coronet project has run into heavy waters because of the lukewarm if not openly hostile attitude of most European PTTs and their international bodies CEPT and Eutelsat. Eutelsat, a common organization of European PTTs that operates the ECS satellite system, has tried to get rid of Coronet by pushing it into a corner where it obviously doesn't belong.

Eutelsat has repeated several times in the recent past

the case for maintaining a "single regional international telecommunications satellite system" in Europe.[8] Ever since November 1983 the organization has made it clear that because it is still in its initial stage of operation and its economic viability is still being gradually established, any new satellite system, and consequently GDL (GDL stands for Grand-Duchy of Luxembourg and refers in Eutelsat jargon to the satellite system Coronet is planning to operate), whose mission is to provide "international public telecommunications services" in Europe could not fail to cause "significant economic harm" to the organization.

Eutelsat sticks to an almost all-encompassing definition of what falls under "international public telecommunications services," reflecting the preoccupation of European PTTS to keep as much of their monopoly as they can. At a meeting of the Assembly of Eutelsat Signatory Parties held in Paris in early November 1983 and which dealt extensively with the GDL/Coronet dossier, it was already established that the satellite television distribution service is indeed a public telecommunications service as defined by the Eutelsat Convention and by the Radio Regulations of the International Telecommunication Union. Eutelsat is of the opinion that the market in Europe for satellite communication has not yet developed to the extent where more than one system can be justified. The organization also likes to emphasize that its operational planning is already such as to envisage meeting Luxembourg's requirements for international telecommunications.

Eutelsat believes that by its determination, it is reaffirming the preeminence of a regional satellite system in Europe, operated jointly by twenty telecommunications administrations or entities (including the Luxembourg PTT), whose public service role should be preserved in the interest of the users and to make worthwhile the considerable investments such a system has required. This attitude was strongly reaffirmed at the end of September 1984 when Eutelsat's policy making body, the ECS council gave the green light for the launch of a third ECS satellite (Eutelsat F-3). The three satellite-in-orbit system Eutelsat will have at its disposal will provide several more transponders for TV relay, and also more capacity for the business services system (SMS). This decision was quite clearly a victory for Eutelsat's secretary general

Andrea Caruso, who has been Coronet's most outspoken oppo-
nent. He declared himself confident that his organization will now
be able to meet the transponder demand in Europe, particularly
for TV relay, and he added with a glance at Coronet: "This should
stop certain private initiatives announced in a number of countries
in Europe."[9]

The rift between Coronet and Eutelsat has been limited
so far to serious warnings addressed to Luxembourg. For instance,
Eutelsat's assembly of signatory parties adopted at its sixteenth
meeting in November 1983 a resolution stating that participation
by signatory parties in the use of the GDL networks, with the
consequent extension of their services beyond the Grand-Duchy
of Luxembourg, to provide international public telecommunications services in Europe, will have serious consequences on the
Eutelsat system and the investment and objectives of the CEPT
administrations which are members of Eutelsat. This attitude was
restated at the seventeenth meeting of Eutelsat's assembly of signatory parties (Paris, May 14–17, 1984) and the assembly decided
to "urge all Eutelsat signatories to refrain from entering into any
arrangements which may lead to the establishment and use of
any new satellite systems providing international public telecommunications services in Europe and which might cause potential
harmful competition to Eutelsat."[10]

It must be noted that the secretary general did not
succeed with a resolution that contained much stronger wording
that has been outlined in a document with the title "GDL System
Compatibility with the Eutelsat System.[11] In this document the
secretary general invited the assembly to "conclude that the proposed GDL system and any other European network intending to
provide international public telecommunications services in Europe has to be considered in the same way as the proposed competitors of INTELSAT over the Atlantic basin." The secretary general did not hesitate to deride the Coronet project in order to gain
support for his hostile attitude. In the document he expressed his
personal conclusion in the following way, which in turn provoked
a strong reaction from Luxembourg's prime minister, Pierre Werner: "It is clear that the GDL project is driven by North American
private interests in the spacecraft manufacturing and the distri-

bution of TV material. These interest groups may, via this project, dump their surplus facilities and already available American programmes over Europe."

While this extreme view was not ratified by Eutelsat's bodies, it gave the start to an all-out campaign against the Coronet project and its American coloration masterminded by France. France was, of course, afraid that Luxembourg might drop out of an arrangement between the two countries for the exploitation in common of the French DBS TDF-1 that would allot two out of four transponders to Luxembourg-based commercial broadcaster "Compagnie Luxembourgeoise de Télédiffusion" (CLT/RTL).[12]

At a meeting of the telecommunications commission of CEPT (Conférence Européenne des Postes et Télécommunications) in Montpellier, June 20–27, 1984, the Coronet project was once more a prominent topic. Without taking a binding decision, the telecommunications commission subscribed to the view adopted by another CEPT body the CCTS, at its meeting in The Hague, April 25–27, 1984. Both CEPT bodies take note of the problems caused by the establishment of the GDL system. They also recall that the fourteenth session of the INTELSAT meeting of signatories (Washington, D.C. April 1984) unanimously adopted a resolution on the implications of the development of separate systems on the viability of the INTELSAT system and the economics of its service offering. The two CEPT bodies also restate that this IN-TELSAT resolution invites the signatories not to enter into agreements that might lead to the establishment and subsequent use of separate systems to carry traffic from or to their countries. Both CCTS and the telecommunications commission emphasize that Eutelsat system viability might be seriously jeopardized if one or more systems separate from Eutelsat were established in Europe to carry international telecommunications traffic. Consequently, both bodies invited the CEPT administrations to adopt the same firm attitude vis-à-vis these systems as they adopted at the IN-TELSAT meeting of signatories with respect to the separate transoceanic networks.

Luxembourg feels that these actions on the regulatory front are no serious impediment to the GDL/Coronet plans, the more so since some other major European countries have come to realize that the monopolistic attitude taken by Eutelsat could seriously hamper their own national telecommunications satellite plans. The general feeling is that Coronet was a test case and most countries are in a way happy that Luxembourg offered itself to bear the brunt by testing ways and means to achieve deregulation in Europe. Luxembourg is confident that the winners in the satellite race will be those who provide their customers with the best service at the best price. In view of the significant volume of unsatisfied demand for TV transponder leases in Europe, Coronet is confident it will find a market for its services.

NOTES

1. Brenda Maddox, "In Pursuit of the Unviable," *Connections* August 13, 1984.

2. David Webster, "Direct Broadcast Satellites: Proximity, Sovereignty and National Identity," *Foreign Affairs*, Summer 1984.

3. Commission of the European Communities, *Europe-Wide Television: Green Paper on the Establishment of a Common Market in Broadcasting by Cable and Satellite* (Brussels, 1984), COM (84) 300 final.

4. Cable television in Western Europe: "A Licence to Print Money?" the Economist Intelligence Unit, London 1983; "Cable TV Communication," in Western Europe, CIT Research, London, 1984.

5. Quoted in *Cable & Satellite Europe*, April 1984.

6. Commission of the European Communities, *Realities and Tendencies in European Television: Perspectives and Options* (Brussels, 1983), COM (83), 229 final.

7. Jonathan Miller, "Europe Needs an Open Market", *Cable & Satellite Europe*, June 1984.

8. Cf. Interim Eutelsat, press release No. 27, May 18, 1984.

9. Quoted in *Cable & Satellite Europe*, Oct. 1984.

10. The minutes and decisions of this meeting are reproduced in Eutelsat document APS 17—3E.

11. Eutelsat document APS 17—16E, May 4, 1984.

12. On CLT/RTL and its French connection, sec Mario Hirsch, "Radio-Télé-Luxembourg: Gebremster Vormarsch?" *Rundfunk und Fernsehen*, Feb. 1983.

N I N E

Canada's Space Policy

W. M. EVANS

C anada has been involved in space almost since the dawn of the space age. Throughout this past quarter century our space policies have evolved to reflect the increasing importance and complexity of our space program. They are still evolving, as indeed they must in order to keep up with this rapidly changing, leading-edge technology. What I would like to do in the space available is to trace the development of Canadian space policy and explain the critical factors that have affected it.

SPACE POLICY DEVELOPMENT

Over the years, Canadian space policy has provided the guidance required to pursue two fundamental objectives of the Canadian space program. These objectives have been stated differently at different times, but in essence they have been with us since the inception of the program. They are: (a) to encourage the use of space technology to meet national needs; and (b) to develop an indigenous space operations and manufacturing industry.

Before looking at our policy development, a word or two on our progress in achieving these two fundamental goals. On the applications side, Canada was the first nation to establish a domestic communications system using geostationary satellites, not by accident but because communications is our lifeblood. We are now one of the world's largest users of space systems, meeting national needs in fields as diverse as telecommunications, broadcasting, remote sensing for resource management and surveillance, weather forecasting, search and rescue, and navigation.

In the process of ensuring that Canadians benefit from the application of space technology, we have been able to develop a world-class space industry that in the past few years has been growing at about 50 percent per year. Annual sales have reached almost $300 million (Canadian) with over 70 percent exported. A recent study by the OECD shows that Canada is the only country where the sales of its space industry exceed the space budget of the government. This has been the case since 1978.

We are proud of our accomplishments, particularly in view of our relatively small space budget. This year, the government will spend about $150 million on space. This level of expenditure places Canada eighth in the free world in terms of space expenditures as a percent of GNP.

THE IMPORTANCE OF INTERNATIONAL COOPERATION

International cooperation is a fundamental policy guiding the conduct of our space program. All of our major government space programs have been jointly undertaken in cooperation with other nations. More and more Canadian space industries are entering into teaming arrangements or partnership arrangements with firms in other countries. This emphasis on international cooperation is in recognition of our relatively small domestic market, the relatively small size of our industry, and the very high cost of exploratory space programs.

Our space program has benefited remarkably from this policy of internationalization. Cost sharing has allowed us to undertake more than we otherwise would have. Work sharing has allowed us to concentrate on our areas of expertise. For example, on the L-Sat (or OLYMPUS) program with ESA we are doing the solar arrays and certain communications payload elements. Technology sharing has prevented unnecessary duplication of effort. A good example of this occurred on the Communications Technology Satellite program when NASA supplied us with the 200 watt traveling wave tube. International industrial arrangements have opened up export markets for our expertise and have given us access to the best foreign technology for our own needs.

All of our international joint programs have been very successful. This is because the following criteria were met in each case: (a) a common set of objectives was agreed; (b) program commitment was made at an adequate level; (c) good working relationships were established; and (d) free and open exchange of data occurred.

INDUSTRIAL ASPECTS OF SPACE POLICY

Canada's expertise in space technology developed initially in government laboratories and our first satellite, ALOUETTE I, was designed and built by a team of government scientists and engineers. By the time the satellite was launched in 1962, there was a realization on the part of the government that space technology could be used for practical applications as well as for scientific exploration. The commercial implications were foreseen and a very important policy decision was taken to transfer the technology to Canadian industry. This policy decision was the first major step in encouraging a Canadian industrial space capability.

The mechanism chosen for transfering the technology also had a marked impact on the long-term development of a close government/industry relationship in space technology and applications. Industry personnel came and worked with the government team on ALOUETTE II and then gradually, on our next two satellites, government involvement was reduced until industry took complete responsibility for meeting specifications on ISIS II in the early 1970s. Not only was this an effective technology transfer process, it built the personal relationships necessary for the continued government/industry partnership that is one of the strengths of the Canadian space program.

With the development of an indigenous space industry as a prime objective, subsequent policy decisions have more or less been concerned with ways and means for its implementation. One of the more innovative steps was taken when the government established Telesat Canada. The company's charter requires it to do two things: first, provide telecommunications services on a commercial basis; and second, utilize Canadian industry to the extent practicable, bearing in mind the company's commercial nature. Thus, through the use of offset arrangements and Canadian content provisions, Telesat procurements have been, and will continue to be, vital elements in the development of the Canadian space industry.

The development of a prime contractor for satellites has been the major thrust of our space industry development policies for almost a decade. This policy thrust has not always received whole-hearted support within government or from all segments of the space industry. However, with the recent success demonstrated in the export market, it is now generally agreed that having a prime contractor has been a stimulus to the development of our industry and the achievement of substantial economic benefits.

While the reasons for the prime contractor policy may be specific to the Canadian context, it will be worthwhile to say a few words about why we felt we needed a prime contractor before describing the policies and programs we put in place to implement the policy. By the end of the ALOUETTE/ISIS series of scientific satellites, Canadian industry had demonstrated its capability to design and manufacture experimental satellites for the government. There were expectations that the industry would supply Telesat's communications satellites. The ANIK-A procurement, however, demonstrated that our industry was far from being commercially competitive. We could not match the price, delivery, or performance guaranteed by Hughes Aircraft in their bid to Telesat. The letting of the ANIK-A satellite contract to Hughes Aircraft was a major disappointment to Canadian industry and was cause for an in-depth analysis by the government of its space policy.

At issue was the question of whether it was in our

long-term national interest to have our domestic satellite communications systems supplied by our own industry or to purchase the systems offshore. The factors that were considered included: sovereignty, security of supply, balance of payments, economic development, high quality employment, export potential, and the cost of achieving success. In the end, the government decided that we should continue to develop our industry so that we could meet our own demand for satellite systems.

The ANIK-A procurement had shown that Canadian industry did not have all the technology necessary for commercial communication satellites; that the industrial structure was inappropriate for bidding on commercial systems; and that the industry did not have the facilities required to integrate and environmentally test complete satellites. In short, while our industry had strength in various subsystem areas, it lacked the structure and capability to be a prime contractor. If we were to meet the government's expectations for supplying our own needs, we would have to have a prime contractor.

The objective was clear. We also clearly wanted to move forward in the development of technology and in establishing the necessary conditions and facilities for applying space technology. Policies and programs were then put in place to address each of these. On the technological front, it was realized that we should address the up and coming technologies that were likely to be required five to ten years down the road rather than attempt to catch up with current technology. The Communications Technology Satellite program (later called HERMES) was initiated to develop $12/14$ GH$_z$ (or Ku-band) communications payloads, to develop experience with 3-axis stabilization, and to develop large flexible solar arrays. The program was jointly undertaken with NASA and we were able to tap that organization's immense pool of technology. When launched, HERMES was the most powerful communications satellite in the world, and the first to operate in the new frequency band.

The satellite also paved the way for Telesat to provide commercial service in this band. Their next satellite, ANIK-B, included a Ku-band transponder designed and built in Canada using technology developed on HERMES.

The second major program was also undertaken during this period. The Remote Manipulator System project, or CANADARM as it is now called, was initiated as Canada's contribution to the U.S. shuttle program. Our objective was to continue the technological development of our industry in a specific area of specialization where industry felt it could exploit the technology commercially. The program also offered the distinct advantage of giving us first-hand experience with the complete shuttle system—an experience which would be invaluable in future space programs.

As an aside, it should be noted that the CANADARM project, because it was so clearly associated with the manned space program, has captured the imagination and interest of the general public to an extent far beyond anything else we have done in space. This interest provided the necessary base of public support for the decision for Canada to enter into manned space activities with the establishment of the Canadian Astronaut Program. Our first astronaut flew in October.

Technology and industrial development was also a primary thrust behind Canada's decision to seek a cooperative agreement with the European Space Agency. Through this agreement, Canadian industry has been able to participate in and benefit from R&D work sponsored by the agency. Our participation in major ESA projects such as L-Sat, now called OLYMPUS, has enabled Canadian industry to continue to develop their technological excellence in some of their areas of expertise (in this case solar arrays and communications payload equipment) and to develop close industrial linkages with European firms. We expect these linkages will become of strategic importance as the space industry around the world matures.

We have recently renewed our cooperative agreement with ESA as a result of the positive industrial benefits we achieved during our first five-year agreement. Our participation in the general budget of the agency has increased in recognition of our increased overall involvement with the agency.

To address the facilities question, it was decided that the government would invest in a national facility for the integration and environmental testing of satellites. The first phase of the facility (now called the David Florida Laboratory) was constructed for the HERMES program. It has since been expanded once and is now undergoing a further expansion. The government owns and operates this world-class testing facility and industry is charged a fee for its use. This arrangement, where the government makes an up-front investment because industry is unable to do so and then charges a fee for use, has been a very important element in the development of our space industry.

The structural problem was more difficult to address. No single Canadian company had the expertise to be a prime contractor. However, in theory we did have the necessary expertise spread throughout several companies. The initiative for restructuring had to come from industry, a fact recognized by the industry. They addressed the problem several times, looking at various consortia or teaming arrangements, but none of these attempts at rationalization was successful. Eventually, one company, SPAR Aerospace Ltd., undertook to purchase the space expertise located in two of the other major companies. This gave SPAR the capability to be a prime contractor.

With the prime contractor capability in place, supported by a strong subsystem and component industry, and with the necessary space facilities established, government policies in support of the space industry could now be directed toward the longer-term issue of ensuring commercial success for the industry in both the domestic and export markets. This is the phase that we are now in. The government has put into place a number of policies and programs aimed at taking advantage of our domestic requirements, both commercial and government, to further develop the technical and commercial capabilities of our industry and to ensure a healthy export business.

Perhaps one of the more important policies was the joint government/industry decision to define the prime contractor policy more precisely as the "limited prime contractor policy." By this, we mean that our prime would not have to have an in-house capability to design and produce all major subsystems. Instead, SPAR would be encouraged to develop commercial partnership arrangements with foreign companies to gain access to specific technologies (for example, the space bus or platform) and to enhance the export market potential for our own technologies. To cement these relationships, SPAR would be able to bring to the table its own technologies and enhanced access to the Canadian domestic market. So far, two of these relationships have been established to the benefit of all concerned.

Our technology support policy has been augmented to ensure that the R&D undertaken can be driven by market opportunities perceived by industry as well as by the mission needs of the government. The support programs put in place recognize the main beneficiary of the program so that the more industry benefits, the less is the government's share of the funding.

Support for export marketing follows normal government policies and includes, as necessary, our trade commission service, export financing, aid programs, technology transfer agreements, training agreements, and ministerial support.

THE MSAT PROGRAM

The proposed MSAT program provides a good example of government/industry cooperation. It is also a good example of how international cooperation can assist in the development of new space-based services even when there will be intense industrial competition for the supply of hardware.

Since 1980, NASA and Canada have been studying the use of satellites in the 806-890 MHZ band to improve mobile communications in both countries. These studies have shown that a satellite system could provide cost effective mobile radio and telephone services to a variety of low-cost mobile terminals (land, sea, and air). These studies have shown that the nature of mobile communications via satellite is such that regional frequency and orbital slot coordination/sharing is fundamental to success. In addition, there are considerable benefits to be obtained if the United States and Canada cooperate to obtain systems and equipment compatibility, joint procurement of the space segment hardware, service restoral arrangements for space and ground facilities, sharing of nonrecovery costs, and the sale of unused capacity on an interim basis between the systems of the two countries. Thus, in

1983 NASA and the Canadian Department of Communications (DOC) signed an agreement for cooperation in the definition of a space communications program leading to the joint development of a mobile satellite communications system to meet the needs of both countries. Because Canadian and American commercial interests are willing to become involved in the establishment of a commercial MSAT system, the arrangement is structured to encourage cooperation not only between governments but also to involve appropriate commercial telecommunications carriers in the two countries.

The Mobile Satellite Communications cooperative program envisioned by the arrangement would have the following objectives:

1. provide an operational commercial mobile-satellite communications capability to the United States and Canada;
2. provide an experimental and test capability for both countries;
3. foster the development of mobile satellite technology, systems, and services;
4. develop and evaluate system networking techniques, mobile terminal technology, space segment technology, and mobile satellite systems and services in the 806-890 MHZ band designed to provide ubiquitous nationwide mobile radio, telephone, and data communications in North America;
5. characterize systems and link performance in an operational environment;
6. develop and evaluate frequency reuse and spectrum-efficient techniques;
7. identify frequency sharing and interference criteria to help define the regulatory framework for commercial satellite services.

Both NASA and DOC are encouraging U.S. and Canadian telecommunications carriers to work together to design and procure the space segment to meet government requirements as well as their own, to initiate early commercial satellite service, and to secure appropriate frequency allocations from their respective governments.

As a result of this cooperative arrangement, it is foreseen that the first mobile satellite system would involve investments by four parties: the U.S. government, the Canadian government, the U.S. private sector, and the Canadian private sector. Such extensive cooperation in an area of the commercial use of space is probably unique.

Within this framework of cooperation, however, both governments are initiating the development of technologies by their respective manufacturing industries to ensure that both countries share in the industrial benefits that could accrue from building the system. The projected market for mobile satellite ground terminals is substantial and it is expected that the competition of supplying these terminals will be intense.

For the space segment, there will be substantial savings if a common procurement can be achieved. This face should encourage cooperation between Canadian and U.S. industries to ensure the equitable sharing of space segment hardware contracts. If this is accomplished, it could lead to joint exploitation of the international market for similar systems.

CONCLUSION

Canada has benefited substantially from the atmosphere of international cooperation that has characterized space research and development in the past. We believe our partners have equally benefited. As a result of this cooperation, nations such as Canada have been able to participate in and contribute to a number of space activities in such areas as communications, remote sensing, science, and search and rescue.

Our experience with communications satellites has shown that even in an intensely competitive commercial area it is still possible to undertake cooperative programs while at the same time encouraging the industries of the partners to achieve their business objectives. It has been shown that cooperation and competition can be compatible objectives in the conduct of space activities. It is our hope that as we rush toward the commercialization of space in many new and promising areas such as remote sensing and space-based manufacturing, our experiences in the field of communications are remembered. International cooperation can play a vital role in furthering the development and use of space technology and can, in fact, lead to constructive competition for the supply of space hardware and associated services. We are looking forward to the challenges of the future with the expectation that cooperation will continue to be a significant factor in the space programs of the world.

T E N

Research and Development Policy in the United States: Implications for Satellite Communications

FRED W. WEINGARTEN AND CHARLES WILK

Innovation in electronic information technology has transformed U.S. telecommunications over the last decades. Research and development (R&D) in fields as diverse as microelectronics, space vehicles, and optics have transformed the basic technology of transmitting information. This transformation has also helped to profoundly affect both the structure of the communication and information services industry and the way in which government regulates those industries. Now these domestic changes are beginning to create serious strains in a world system that has, by and large, not adapted as drastically.

The resulting international policy debate blends together a vast array of issues, from technical standards and resources allocation to the free flow of information and First Amendment principles. Throughout, there is an unavoidable tension between the pragmatic desire to achieve a competitive position in the communications market, and concerns over maintaining cultural diversity and national sovereignty as well as perceived inequities in the consumption of scarce resources.[1]

Questions about technology, where it is and where it is going, weave through this debate. Technology is seen, at least in part, as the cause of some of these problems, and in other cases, as a possible solution. Hence, thoughtful policymakers in most countries look at research and development with a curious binocular perspective. They believe that it is in the national interest to pursue R&D in information technology aggressively. At the same time, they must also wonder what new problems technologies now in the laboratory hold for the future. Certainly that dual perspective exists in the area of satellite communications. This paper addresses aspects of the Federal policy role in technology, and touches on a few of the current concerns facing the U.S. Congress and such international bodies as the International Telecommunication Union, UNESCO, and the General Agreement on Tariffs and Trade (GATT).

To address some of these concerns, the Office of Technology Assessment was asked by the Congress to examine the state of R&D in information technology. The report of that assessment was released in February 1985.[2] In addition, OTA has released a report on cooperation and competition in space.[3] These reports will be the basis for much of the following discussion.

THE IMPORTANCE OF GOVERNMENT R&D FOR SATELLITES

An understanding of U.S. R&D policy as it relates to satellites is important for two reasons. First, the range of R&D that relates in some way to satellite communications is diverse. Improvements in future systems will depend on fundamental advances in a variety of such scientific and technical fields as microelectronics, antenna design, and the evolving mathematics of digital signal encoding, as well as on a more comprehensive understanding of radio wave propagation. The future of satellites will depend, in a reverse way, on how fast alternate technologies, such as fiber optics, advance. There are a number of application technologies that will affect the demand for satellite communication services. Will future demands for capacity be dominated simply by the need for more telephone channels or, as may be more likely, for more sophisticated services, such as videoconferencing, high definition broadcast television, mobile services, or some other application—such as interactive graphics? Answers to these questions depend on the results of research not only in hard science and engineering, but in such fields as social science, psychology, management, and the regulatory environment.

In addition, the Federal government has a history of involvement with R&D in the broad sense that reaches back nearly two centuries. How it perceives its responsibility with respect to the private sector to help advance satellite communication technology, and what action it takes, are deeply affected by this history and by firmly established science and technology policies. Although these policies do evolve over time, they are rarely altered drastically to serve the needs of just one technology.

We pause to note that, for obvious reasons, the Defense Department has a particularly keen interest in communications, including a continuing interest in satellite technology. But much of DOD R&D in this area is and always has been highly classified. This paper does not take that work into account except to acknowledge its existence and to suggest that there is probably a limited and slow, but also unidentifiable and thus incalculable, flow of technology from those defense efforts into the civilian sector.

INFORMATION TECHNOLOGY R&D

Science and technology policy in the United States generally distinguishes three broad groups of participants: government, industry, and nonprofit institutions—most notably, universities, although some new experimental forms are evolving. This section briefly focuses on industry and universities and then dwells at some greater length on government actions and policies.

Industry

Estimates of industry R&D levels are particularly hard to come by. Funding levels are often considered to be proprietary, definitions vary, and investments by small, entrepreneurial firms are hard to measure. We estimated that all U.S. industrial investments in information technology R&D amounted to approximately $10 billion in 1982. The industry is broadly based, in general, and sustains a relatively high level of R&D, ranging from new product development performed by the very small firms to a wide mixture of basic research and development performed in large industrial laboratories.

The most prominent example of a large laboratory is AT&T's Bell Laboratories, one of the nation's first industrial laboratories, and still one of the largest and, by many accounts, most productive. In fact, Bell Telephone Laboratories was the only industrial center able to invest in significant R&D in satellite communications during the early, high risk years when this technology was first being developed. Lately, many science policy experts have been expressing concern about the fate of Bell Labs in light of the divestiture of AT&T and the deregulation of the telephone industry. They argue that divestiture may deprive AT&T of the ready source of funds from the local operating companies for fundamental research, and that deregulation may pressure the labs to increase significantly its emphasis on short-term development at the expense of longer-term applied and basic research. OTA's investigation suggests that there is little evidence or economic theory that would support such a bleak scenario over the short term, but that longer-term prospects were less clear.

A new trend among firms in the information technology industry is the formation of joint research ventures. The best known of these is probably the Microelectronics and Computer Technology Corporation (MCC), conceived by Bill Norris of Control Data as a U.S. response to the challenge of joint technology programs promoted by the Japanese government. Other such centers include the Semiconductor Research Consortium (SRC), that principally offers basic research grants to investigators in university laboratories, and the Microelectronics Center of North Carolina (MCNC), that combines both industrial and state funding. Out of concern about possible antitrust violations, most of these ventures have been carefully designed to concentrate on nonappropriable basic or applied research. Many are associated with or provide support to university research projects.

Universities

Universities are another important institutional element in the U.S. R&D picture, principally as performers of basic research. Much of it is funded by the Federal government, but an increasing amount of private research funds are also being channeled to university labs.

The role that universities play in computer and communications research has shifted over the years. In computers, for example, university laboratories were the focus of much pioneering early research. Then, as the development of new generation computers became expensive and difficult to manage, industry took over computer and software design, leaving the university researchers to work on more theoretical problems. Lately, very large scale integrated circuit technology has allowed architecture research to move back on campus, funded by both Federal agencies and private industry. In the past few years, industry and the Federal and State governments have been strengthening their ties with university research centers across the nation, in recognition of the important contributions expected from these centers.

Government

Traditionally, when talking about R&D, the term "government" referred to the Federal government. Recently, however, some State and local governments, assuming a close link between high technology and economic development, have become more active in establishing new research centers and are promoting and funding R&D.[4] The following discussion focuses principally on Federal policy.

In establishing science and technology policies, the U.S. government is pursuing several goals, some of which may be in conflict. Many of these goals relate directly to support of R&D in satellite communications. For example, there is the clear goal of strengthening national defense. A modern military force is critically dependent on worldwide electronic communications for purposes including intelligence and surveillance, missile guidance, navigation, command and control, and so on. The implications for satellite communications, as well as for information technologies in general, is demonstrated by the fact that the Department of Defense accounts for nearly 80 percent of all R&D funding in these areas. The DOD provided the principal funding for the early development of satellite technology, and still provides significant support.

Moreover, advanced communications technology is of established and growing significance in providing for social needs. Communications is more than merely a set of services in the marketplace. U.S. policy reflects the view that communications is a fundamental infrastructure of society and that access to communications is a basic necessity. A major goal of U.S. communications regulation has always been "universal service." One of the motivations behind the formation of INTELSAT was to provide access to satellite telecommunications for the developing world. Communications technology will increasingly form the basis for the delivery of such important public services as education, agriculture, and public health with incalculable benefits to mankind. There are numerous and increasing examples of this in such countries as the United States, Canada, the United Kingdom, India, Indonesia, and Brazil.

The government's R&D is also aimed at stimulating economic growth. One of the key objectives of many NASA programs is to reduce the risks associated with commercializing technology and providing new services. Some examples include the NASA programs in fixed service satellite communications, remote sensing, broadcast satellites, and, for the past decade, mobile satellite communications.

Clearly, over the last two decades the computer and communication industries have grown in importance as components of the economy. These industries have been recognized increasingly by the OECD, the ITU, the World Bank, and other organizations as central to infrastructure development and economic growth. Moreover, they form the basis for a rapidly growing information services industry, and this leverage makes their economic importance to nations even greater. It is not surprising, then, to see national programs to stimulate and support these industries, including satellite communications, cropping up all over the world.

Research in computer and communication science has made major contributions to our understanding of the world around us. Examples range from contributions in artificial intelligence to cognitive psychology to the Nobel Prize winning work of Arno Penzias on the origins of the universe. NASA has always needed to balance its priorities between scientific research and other objectives of the space program.

Computers and high-speed data communication links between computers have also become a vital tool for the conduct of scientific research. The ARPANET has been indispensable to DOD and civil agency supported computer science researchers for many years. Key to the new supercomputer program at NSF will be the development of data communications networks to link the centers together and facilitate researcher access to them.

Certainly a major motivation for the space program has been one of national prestige. Similarly, investments of billions of dollars on such major research instruments as accelerators or telescopes are justified by the argument that international economic and political advantages accrue simply from the world perception of the United States as a technological and scientific leader. The pioneering work in satellite communications clearly has contributed to U.S. national prestige.

The U.S. government also supports R&D necessary to performing the missions of its agencies. One such mission, of course, is national defense, and that has already been discussed. However, many civilian agencies—such as the Social Security Administration and the Treasury Department—make extensive use of information technology. As the administration attempts to reduce Federal spending, and the appropriateness of some civilian agency programs is under question, one might expect civilian support of technology development (in contrast to basic research) to decrease, and such has been the case. One result is that many demonstration programs have been cut in recent years. Even so, for example, the Department of Energy, and to a limited extent

NASA, support development of advanced computer technology, and the Department of Education supports R&D in the educational use of an assortment of information technologies.

Finally, NASA's satellite communications programs have been helpful in promoting U.S. foreign relations. For example, the ATS-1, -3, and -6 satellites have been used extensively throughout portions of the world for applications in public health, safety, and education, and for experiments leading toward commercial applications. Some of the participants included more than a dozen island nations in the Pacific Basin and the Caribbean, as well as India and Canada.

THE PATTERN OF GOVERNMENT FUNDING
OF R&D IN INFORMATION TECHNOLOGY

The Federal government has had a long history of funding R&D in information technology-related fields. It is currently the major sponsor of R&D in the kind of information technology in which it has special interests. It includes artificial intelligence, supercomputers, software engineering, and very large scale integrated circuits (VLSI), all areas in their technological infancy and with enormous potential for military as well as commercial applications. There is a long list of related technologies that have been stimulated by government—often defense or other mission agencies—sponsorship of R&D including radar, guidance system, and satellite communications.

There are some historic examples of intensive government sponsorship of technological development in areas where the potential benefit was expected to be great, but the risks and costs of research were high and therefore unattractive to industry, e.g., computers, aviation, and communications satellites. One of the classic illustrations of a successful, major government contribution to information technology R&D is in the field of satellite communications. The National Aeronautics and Space Administration (NASA) (which currently accounts for about 7 percent of the Federal R&D budget) had the leading role in pioneering technological progress toward commercial development, accelerating the time frame for the introduction of this technology, influencing the structure of the U.S. domestic and international telecommunications common carrier industries, and effecting significant cost savings over the long run.[5]

In these cases, the government, through the undertaking of a number of risky and expensive R&D programs and with extensive private sector involvement, developed a large pool of baseline technology that served to prove the feasibility of geostationary satellite communications. These R&D programs were for the purpose of proving the feasibility of various technological advances, such as geostationary orbiting satellites, electromagnetic propagation of signals from outer space, traveling wave tubes, automatic station keeping, and aircraft communications. The NASA programs initiated to undertake the extensive R&D included the SCORE, ECHO, and RELAY programs, the SYNCOM series of launches that paved the way for INTELSAT I, the first commercial communications satellite, and the Applications Technology Satellite series. The costs for the RELAY, ECHO, and SYNCOM programs alone through 1965 were over $128 million—an amount that few companies could—or would—commit, particularly considering that the feasibility of synchronous satellite operation was seriously questioned.

It is also interesting to note that these NASA programs likely had some important side effects on the structure of the U.S. international satellite communications industry. Because AT&T was the only private company to have heavily invested its own funds for satellite communications R&D—with focus on the nonsynchronous TELSTAR system—it is likely that AT&T would have dominated the new international and domestic satellite communications services industry. Instead, the NASA programs, through continuous transfer of technology to, and close interaction with, commercial firms stimulated the competition that followed the 1972 Federal Communications Commission's decision allowing open entry into the domestic satellite communications services industry.

The market for the supply of satellite communications equipment was also open to competition because of the expertise of contractors. In addition, the international satellite network that evolved is owned and operated by INTELSAT, an international consortium, with the U.S. portion owned and operated by COM-SAT, a broadly based private public corporation.

Other Forms of Government

The Federal government has broad involvement in R&D, as a supporter of different programs, as a performer in government laboratories, and as a consumer of new technological products, for example, supercomputers, and, hence, is a force in stimulating innovation.

The picture over all science and technology is impressive. The Federal government is estimated to fund roughly 50 percent of all R&D carried out in the United States. This estimate does not include indirect incentives and subsidies such as tax credits for R&D.[6] Additionally, the government performs about $11 billion worth of R&D in its own laboratories, and supports over 65 percent of R&D performed on college campuses.

In the case of electronic equipment and communication technology, the proportion of overall Federal support rises to two-thirds. According to the National Science Foundation, Federal basic research support in computer science and electrical engineering was over $200 million in 1984, and applied research in the same fields amounted to over $700 million.[7]

The Department of Defense (DOD) is the principal source of this support. It accounts for about 60 percent of the overall Federal R&D budget. In comparison, NSF accounts for less than 3 percent. Of course, these numbers are skewed by the relatively heavy applied research and development emphasis of DOD expenditures compared to basic research in which NSF clearly plays a much more important role.

The government, through DOD and NASA, has played a major role in the development of satellite communications—particularly geostationary satellite technology.[8] Federally supported R&D through DOD and NASA created, in large part, the foundations of the domestic and international satellite communications industry.

While direct support of R&D is important, the government also affects R&D in several indirect ways, for example, through regulation. For many years, basic research at Bell Labs was supported through AT&T, Western Electric, and Bell Operating Companies. The pattern of research at the laboratories continues to be strongly influenced by FCC decisions deregulating the industry and by court decisions concerning the divestiture of AT&T. Some Bell labs research teams have been broken up and researchers have migrated to different organizations within AT&T or the Bell Operating Company structure. The guaranteed support of research by local companies has been eliminated. Although the effects, good or bad, of these changes are not yet understood, clearly vast changes have been wrought in the world's largest industrial laboratory by virtue of Federal deregulation and application of antitrust law by the Federal courts.

Another way in which the government affects R&D is through intellectual property law. Rooted in the Constitution, intellectual property law (e.g., patents, copyrights, and trademarks) is specifically intended to encourage technological invention and scientific and artistic creation. These days, new technology, particularly information technology, seems to be moving out beyond the reach of traditional intellectual property law.[9] Some in the industry argue that, without expanded protection for the results of R&D, incentives to innovate will erode.

Another intellectual property policy has been the assignment of patents on technology developed with government funds. Public Law 96-517 passed in 1980 was designed to ease the transfer of such patent rights to small business and universities. This policy has been further liberalized by executive order and revised Federal regulations.

Taxes are another way the Federal government influences investments in R&D. The R&D investment tax credit and R&D limited partnerships are policies specifically designed to provide incentives for investment in R&D. The investment credits program will terminate in 1986 unless renewed by Congress, and debate has already started over whether they have been effective.

Export controls are another Federal policy with a potentially significant impact on R&D, particularly in areas like satellite communications that are considered to be sensitive because they have military applications. They affect private sector R&D in two ways. First, controls over the publication and transfer of technical information, while possibly depriving foreign governments access, also can impede its flow within the U.S. technical community. Second, the possibility of limitations of exports can lessen incentives for firms to develop new technology by limiting potential international markets for their products.

SOME POLICY ISSUES

The Federal involvement in R&D in satellite communications raises a number of policy issues, almost all of which are international in character, and present serious challenges to the United States and other countries.

Private Versus Public Investment

NASA's early investment in the development of satellite communications was made when market prospects for the industry were highly uncertain and costs of R&D greater than virtually any private firms could bear.

Now, however, we have an established market for fixed and maritime satellite services that has proved itself successful. To serve that market, we now have a viable domestic satellite industry—manufacturers and service providers. And soon we will have broadcast satellites that send signals directly to residences, and satellites for mobile communications.

To what extent should the government continue to directly fund development of advanced satellite technology? The answer to that question depends, not only on political values and on an assessment of potential markets, but also on the growing threat from foreign nations whose governments do not engage in such political introspection. There has been a U.S. national space policy since 1982 committing the Federal government to encouraging private sector development of space for commercial applications. The government's role, through NASA, is to help reduce some of the technical, financial, and other deterrents to private sector investment.

Developing World

Third World countries have become more demanding of guarantees that the limited natural resources, e.g., electromagnetic spectrum and satellite orbit positions, will be available for their eventual use. There are also continuing concerns about sovereignty issues in international organizations such as the ITU and UNESCO. Developing countries are trying to find ways to deal with these perceived needs and fears, and at the same time to continue the rapid implementation of new applications. These types of issues are likely to continue to emerge as the technology evolves and as demands grow for using the limited natural resources.

Competition for INTELSAT Services

The role of INTELSAT, as initially envisioned, has been eroded by competition in recent years and is threatened further. One source of the erosion is that the number of regional and national satellite communications systems continues to grow worldwide, with each representing foregone traffic for INTELSAT. And now there is the prospect of new businesses vying for a share of the heavy traffic North Atlantic route. Questions are raised concerning how goals of economic efficiency, open competition, and deregulation can be attained while assuring a stable, future role for INTELSAT. The White House recently announced its decision to permit private satellite carriers to provide limited competitive services (video and digital data communications), but not to compete for public switched network services. A second source of erosion to INTELSAT's position is in the form of prospective competition from another technology—fiber optic cable—for transoceanic routes.

International Trade in Equipment and Services

The U.S. trade balance in telecommunications equipment—a field in which the United States has had a tradition of technological leadership—shifted from a $39 million deficit in

1983 to an estimated deficit of $500 million in 1984.[10] The main causes are seen as the growing imports into a unilaterally open U.S. market and much slower growth in U.S. exports, because other countries follow restrictive trade policies. Anticipation of the magnitude of the trade deficit prompted the last Congress to consider legislation that would authorize the President to take unilateral action against countries with restrictive import policies.[11] Executive branch efforts are underway to open foreign markets and to expand the terms of the GATT negotiations to include services and government procurement of telecommunications equipment.

The foregoing attempts to illustrate a number of points. One is that a number of factors are closely intertwined and are helping to shape the current menu of policy issues, including rapidly advancing technology and Federal regulation of industry and its direct and indirect support of R&D. A second point is that domestic policy can have significant implications internationally. And, finally, growing demands for scarce natural resources are causing tensions in international forums.

NOTES

1. Kenneth Leeson, *Computerization: National Strategies, International Effects* (Washington, D. C.: 1983).

2. U.S. Congress, Office of Technology Assessment, *Information Technology and R&D: Critical Trends and Issues* (Washington, D.C.: OTA-CIT-268, February 1985).

3. U.S. Congress, Office of Technology Assessment, *Cooperation and Competition in Space* (Washington, D.C.: GPO, July 1984).

4. U.S. Congress, Office of Technology Assessment, *Technology, Innovation, and Regional Economic Development* (Washington, D.C.: GPO, July 1984).

5. Morris Teubal and Edward Steinmuller, *Government Policy, Innovation, and Economic Growth: Lessons from a Study of Satellite Communications*, pp. 27–287 Research Policy 11 (New York: North Holland, 1982).

6. U.S. Congress, Congressional Budget Office, *Federal Support for R&D Innovation* (Washington, D.C.: GPO, April 1984).

7. National Science Foundation, "Early Release of Summary Statistics of Academic Science and Engineering," December 1983.

8. Morris Teubal and Edward Steinmuller, *Government Policy, Innovation, and Economic Growth.*

9. Fred Weingarten, "Testimony before the House Subcommittee on the Courts Civil Liberties, and the Administration of Justice," U.S. House of Representatives, April 1983.

10. Unpublished data from the Office of the U.S. Special Trade Representative, November 1984.

11. "Telecommunications Trade Act of 1984," S. 2618.

E L E V E N

Commercial Space Policy: Theory and Practice

JERRY FREIBAUM

N ASA's commercial space policy is designed to encourage private involvement in commercial endeavors in space. The policy introduces approaches and incentives to reduce technical, financial, and institutional risks inherent in commercial space ventures to levels competitive with conventional investments. The policy is implemented through such special initiatives as joint endeavor agreements. An application of this type of agreement is described as it pertains to facilitating the birth of a new multibillion dollar communications industry. This is made possible by the development of mobile communications by satellite. This unique capability is expected to provide two-way voice, data, paging, and position location services to mobile users primarily in nonurban areas on a nationwide or regional basis. Thin route inexpensive fixed service telephony may also be possible. This is the culmination of almost ten years of regulatory, technical, experimental, financial, and institutional studies.

SUMMARY OF SPACE COMMUNICATION POLICY: GOALS AND PRINCIPLES

The primary goal of NASA's commercial space policy is to encourage and stimulate free enterprise in space.[1] Implementation of this policy is guided by five principles:

1. *The government should reach out to and establish new links with the private sector.* NASA will broaden its traditional links with the aerospace industry and the science community to include relationships with major non-aerospace firms, new entrepreneurial ventures, as well as the financial and academic communities.

2. *Regardless of the government's view of a project's feasibility, it should not impede private efforts to undertake commercial space ventures.* If the private sector is willing to make the necessary investment, the project's feasibility should be allowed to be determined by the marketplace and the creativity of the entrepreneur rather than the government's opinion of its viability.

3. *If the private sector can operate a space venture more efficiently than the government, then such commercialization should be encouraged.* When developing new public space programs, the government should actively consider the view of, and the potential effect on, private venture.

4. *The government should invest in high-leverage research and space facilities which encourage private investment. However, the government should not expend tax dollars for endeavors the private sector is willing to underwrite.* This will provide at least two benefits. First, it will enable NASA to concentrate a greater percentage of its resources on advancing the technological state-of-the-art in areas where the investment is too great for the private sector. Second, it will engage the private sector's applications and marketing skills for getting space benefits to the people.

5. *When a significant government contribution to a commercial endeavor is requested, two requirements must be met.* First, the private sector *must have significant capital at risk, and second, there must be significant potential benefits for the nation.* In appraising the potential benefits from and determining appropriate government contributions to commercial space proposals, NASA will use an equitable, consistent review process.

A possible exception to these principles would be a commercial venture intended to replace a service or displace a NASA R&D program and/or technology development program of paramount public importance now provided by the government. In that case, the government might require additional prerequisites before commercialization.

IMPLEMENTATION

In implementing this policy, NASA will take an active role in supporting commercial space ventures in the following categories, listed in order of importance:

- new commercial high-technology ventures
- new commercial applications of existing space technology
- commercial ventures resulting from the transfer of existing space programs to the private sector.

NASA will implement initiatives to reduce the technical, financial and institutional risks associated with doing business in space.

- To reduce technical risks, NASA will support research aimed at commercial applications; ease access to NASA experimental facilities; establish scheduled flight opportunities for commercial payloads; expand the availability of space technology information of commercial interest, and support the development of facilities necessary for commercial uses of space.
- To reduce financial risks, NASA will continue to offer reduced-rate space transportation for high-technology space endeavors; assist in integrating commercial equipment with the shuttle; provide seed-funding to stimulate commercial space ventures; and, under certain circumstances, purchase commercial space products and services and offer some exclusivity.

- To reduce institutional risks, NASA will speed integration of commercial payloads into the Orbiter; shorten proposal evaluation time for NASA/private sector joint endeavor proposals; establish procedures to encourage development of space hardware and services with private capital instead of government funds; and introduce new institutional approaches for strengthening NASA's support of private investment in space.

A high-level Commercial Space Office has been formed within NASA as a focal point for commercial space matters. This office is responsible for implementing the NASA policy to stimulate space commerce and has sufficient authority and resources to fully carry out this assignment.

MOBILE COMMUNICATIONS VIA SATELLITE —A NASA INITIATIVE

A satellite-relayed communications system can provide voice and data communications services to mobile users throughout the western hemisphere.[2] Its operation would be similar to that of terrestrial-based land mobile communications systems in which vehicles within line-of-sight of a 50–400 foot ground-based relay tower can communicate with one another. The lines-of-sight of these terrestrial-based systems range only from 3–40 miles. A satellite in geostationary orbit acts as a 22,000 mile high "relay tower," extending the communications system lines-of-sight over almost all of the western hemisphere. Using this height advantage to provide line-of-sight to remote and/or thinly populated areas, mobile communications via satellite can augment and extend terrestrial based mobile communications systems which primarily serve urban areas.

The proposed new service would help the United States achieve affordable nationwide mobile communications by extending existing and planned mobile telephone and private mobile radio services into rural areas not currently served.

User Equipment Would Be Simple, Affordable, Reliable, Small
The intent is ultimately to use the same mobile communications equipment for terrestrial and space systems.[3] Space system users would be able to access terrestrial networks and vice versa. The cost of satellite-compatible mobile communications equipment would be similar to the price of existing terrestrial mobile communications equipment ($500–$2,500). Applicable user charges are expected to be comparable to terrestrial telephone services or private networks.

There Is a National Need
Rural or non-Metropolitan Statistical Areas (MSA) represent 85 percent of the geographical area of the United States, 25 percent of the population (60 million people) and 20 percent (15 million workers) of the nation's commercial/industrial activity. The characteristics of commercial and local government activities and employment are essentially the same within and outside the MSA's. Nationwide, 27 percent of local government activity is located outside the MSA's (e.g., fire, police, school bus, emergency).
There is a recognized need for mobile communications in rural areas. However, terrestrial cellular systems are not expected to expand significantly into rural areas, and terrestrial private radio networks may be economically and/or geographically impractical in many rural areas. Since satellites are well suited for wide area, dispersed population coverage mobile communications via satellite augmenting terrestrial services may be much better able to serve widely dispersed and rural users.
Unique wide area coverage and reliability requirements of many Federal, state, and local agencies, and other public safety organizations can be met only through use of a space service that has a mobile communications capability. The Congress has recently imposed a requirement for nationwide continuity of mobile communications for public safety purposes. There are similar needs in the private sector. Only mobile communications via satellite augmenting (not competing with) terrestrial mobile communications can provide nationwide coverage. And finally, the results of ten years of user experiments and studies involving

industries with widely diverse and rural operations (e.g., trucking, oil and gas, electric power utilities, rural telephone, and emergency medical services) demonstrate the need.[4]

It Is Technically Feasible
The present state-of-the-art is adequate for a first generation system. NASA's experimental ATS-6, in 1974, demonstrated the ability of a 10-meter satellite antenna to support terrestrial mobile units using modified off-the-shelf mobile communications equipment with small omni-directional and medium gain antennas. Canada plans to develop and provide mobile communications via satellite in 1988.[5] U.S. corporations have filed with the FCC for licenses to build satellites and to operate similar systems in the late 1980s.

By the 1990s, a later generation medium to high capacity satellite may use a 20–55-meter antenna with 10–100 spot beams ensuring spectrum conservation through frequency reuse.

It Is Economically Viable
The nonurban market for mobile communications units is estimated to range between 640,000 and 2,445,000 by 1995. Annual revenues to a system operator of between $0.5 and $1.0 billion are projected and internal rates of return of 20 percent to 40 percent are considered realizable.[6] Venture capital has been raised and commercial applications have been filed with the FCC for authority to build and launch satellites for mobile communications and to offer mobile communications via those satellites.

WHERE DO WE GO FROM HERE?

For the United States to begin commercialization, provide for growth, and continue with high risk technology development, adequate frequency spectrum must be allocated by the FCC. This issue is the subject of a current FCC Rulemaking Proceeding, RM-84-1234, which is in response to NASA's November 1982 Petition for Rulemaking.[7]

NASA's Goal And Role

NASA's goal is to facilitate and encourage the commercialization of its technology. This process involves minimizing technology, regulatory, and financial risks. NASA is now at the threshold of culminating nearly ten years of effort toward achieving this goal with respect to mobile communications.[8] Large technology and financial risks have been reduced, leaving the regulatory risk as the primary obstacle in the path of commercialization. In many ways the regulatory process has served to "protect" established institutions but at a heavy cost. This process has suppressed or slowed down the development and the commercialization of new communications technology. Potential new communication services or new technological advances are almost always perceived as a competitive threat to the nonsponsors. New services and technical innovation also represent impending change. The prospect of change—any kind—is always uncomfortable and more often than not somewhat frightening to existing institutions. For years, large well-established U.S. entities opposed the introduction of domestic satellite services back in the sixties. Costly years of regulatory proceedings and lobbying finally cleared the way for U.S. commercial services, but not before Canada charged ahead and obtained three choice geostationary orbit positions at the expense of U.S. national interests. The same process took place for more than ten years with respect to broadcast satellite services. Twelve years of costly proceedings transpired before terrestrial cellular mobile communications became a reality. Proceedings for a Land Mobile Satellite Service actually began in the mid '70s during U.S. preparations for the 1979 World Administrative Radio Conference (WARC). Proposals for enabling the satellite service were introduced by NASA during eight FCC Notices of Inquiry in the 1970s.[9] This persistence, supported by many years of study and experiments and technology development, culminated in a U.S. position advocating primary allocations in the 800 MHz band shared with terrestrial Land Mobile Services. The 1979 WARC approved these proposals.[10] Now, ten years after the initial proposals, these efforts are about to bear fruit domestically.

The Federal Communications Commission released a

Notice of Proposed Rulemaking on January 28, 1985 (General Docket no. 84-1234), in response to NASA's November 1982 Petition for the Establishment of a Land Mobile Satellite Service (RM 4247). Key provisions include:

- A strong statement that the Mobile Satellite Service is needed;
- A proposed primary allocation of 821–825 MHz;
- A statement of intent to allocate additional bandwidth in the 1500/1600 MHz band ("L" band)
- A recognition that communication services other than mobile (e.g., fixed, broadcast, data, paging) through the mobile should be permitted provided that mobile is the primary use
- Applications for MSS systems will be accepted in parallel with the rulemaking proceeding.

Comments are due by April 22, 1985, and replies by May 22, 1985. Mobile satellite applications for licenses are due April 30, 1985. Since the extent to which Mobile Satellite Services will be implemented and prove useful is very much a function of the public's involvement in the regulatory process, NASA is bringing an awareness of this proceeding to the public.

- NASA's role and interest in this FCC rule-making proceeding is based upon long-standing precedent. NASA developed critical technology and helped define parameters and frequency allocations for the Fixed, Broadcast, and Earth Resource Satellite Services (among others) well in advance of the definition of institutional roles and responsibilities for these services.
- The spectrum allocation is an essential step for commercial activity to begin. Without the assurance of a primary frequency allocation and adequate bandwidth, commercial mobile communications via satellite would never become a reality. Clearly, it is also not desirable for NASA to invest large sums of R&D dollars in high risk communications technology if

this technology cannot ultimately be exploited by the private and public sectors.

- NASA is prepared to enter into a joint agreement with U.S. industry to facilitate commercialization, develop the technology needed for growth, and to conduct experiments. On February 25, 1985, NASA issued an Opportunity Notice for a Mobile Satellite Agreement in parallel with the FCC rule-making proceeding on MSS allocations. NASA intends to offer the U.S. firm licensed by the FCC to provide Mobile Satellite Services specified standard shuttle launch services in return for a specified amount of satellite channel capacity on the first commercial mobile satellite system. NASA and other government agencies will use this capacity for a period of two years to conduct experiments. Details of this arrangement are described in the referenced document.
- NASA is working with Federal and state government organizations to develop communications experiments using the first U.S. commercial satellite for mobile communications.
- Canada's Department of Communications and NASA have signed an agreement (November 1983) to cooperate toward the establishment of commercial mobile communications via satellite.

SUMMARY

Too often, the process of commercializing new communication satellite technology has been excessively long and costly. Regulatory, institutional, and economic barriers have been key factors in causing this delay. NASA has initiated a focused effort to facilitate the commercialization of space technology through technical, financial, and institutional initiatives. As a result of these efforts a new Land Mobile Satellite Service has been developed and is expected to be in commercial operation by the late 1980s.

This service, which is expected to spawn a new multibillion dollar industry, will help the United States achieve affordable nationwide mobile communication by extending existing and planned mobile telephone and private mobile radio services into rural areas not currently served.

NOTES

1. NASA, Commercial Use of Space Policy, NASA Headquarters Code I, October 29, 1984.

2. NASA, Mobile Communications Via Satellite—The Land Mobile Satellite Service, NASA Headquarters Code EC, December 1984.

3. NASA, Petition for Rulemaking, Amendment of Parts 2, 22, and 25 of the Commission's Rules Relative to Satellite Augmentation of Terrestrial Cellular and Non-Cellular Mobile Communications Systems through Establishment of a Mobile Satellite Service, RM 4247, November 24, 1982. NASA Headquarters Code EC.

4. J. Freibaum, NASA, "Need for, and Financial Feasibility of, Satellite-Aided Land Mobile Communications," IEEE International Conference on Communications, Philadelphia, Pa., Conference Record volume 2 of 3, ref. 7H.1 P.A. Castruccio, ECO systems; C. S. Marantz, Citibank, N.A., J. Freibaum, NASA, June 1982.

5. Canada, DOC, Proposed Spectrum Utilization Policy, Government of Canada, Department of Communications, June 1984.

6. Citibank, NA, "Financial Study for a Satellite Land Mobile Communications System," Combined—Phase A—Industry Analysis Report, and Phase B—Financial Structures Report, Corporate Finance Division, Merchant Banking Group, JPL Purchase Order Number BP-73664 for NASA, December 16, 1981.

7. FCC, Notice of Proposed Rulemaking, Amendment of Parts 2, 22, and 25 of the Commission's Rules to Allocate Spectrum for, and to Establish Other Rules and Policies Pertaining to, the Use of Radio Frequencies in a Land Mobile Satellite Service for the Provision of Various Common Carrier Services, General Docket No. 84-1234, RM-4247, adopted November 21, 1984, released January 8, 1985.

8. NASA, Opportunity Notice for a Mobile Satellite Agreement—NASA Headquarters Code EC, February 25, 1985.

9. NASA, "Petition for Reconsideration" before the Federal Communications Commission. "An Inquiry Into the Use of the Bands 825–845 MHz and 870–890 MHz for Cellular Communications Systems," FCC Docket No. 79-318 and Other NASA Filings Under This Docket and Dockets 79-112, 79-113, 80-10 and 80-739, June 22, 1981.

10. International Telecommunication Union, "Final Acts of the World Administrative Radio Conference, Geneva, 1979" (WARC-79). A general revision of the International Radio Regulations.

TWELVE

Piracy of Satellite-Transmitted Copyright Material in the Americas: Bane or Boon?

SYLVIA OSPINA

The development of all kinds of electronic technologies in the past two decades—from cable television to videocassette recorders and satellite transmissions—is taking us from an age of the written word to a new era of audiovisual communications. While literacy is still an important factor in assessing the development of a particular society, it is likely that in the near future a society's development will be measured by the number of television and telephone sets at its disposal.[1]

The vast majority of the communications media are under the effective control of a few countries, and with respect to the use of satellites for television broadcasting, the United States is clearly the leader in the field.[2] At the same time that more television programs intended for U.S. audiences are transmitted by satellite, the number of unintended recipients has also increased. The piracy of satellite signals has great implications for the future development of the economy of the "thieving" countries, as well as for the economic protection of the copyright owners of the pirated programs.

The entities involved in "poaching" satellite signals range from individuals to government organizations. Their acts affect not only the owners of the intellectual property that is being used without remuneration but they also have repercussions on international relations and the formulation of foreign policies.

This paper will consider the interrelationship between copyright laws, the various international organizations whose activities are related to communications by satellite, and some of the implications raised by the piracy of satellite signals. The focus will be on the Americas, or what the International Telecommunication Union has designated as "Region 2."

Copyright legislation differs from country to country and is rarely applicable extraterritorially. Because it is difficult to assess how domestic legislation is applied to the issue of satellite signal piracy, the particular provisions and practices of the countries in Region 2 will not be considered.

The use of geostationary satellites for communication purposes was suggested by Arthur Clarke in 1945,[3] and what seemed to be a science fiction proposal at the time became a reality by the 1960s. The first satellite to be launched was the Russian "Sputnik" in 1957, followed by the launching of an American satellite in 1958. In 1963, the first transatlantic color-television pictures were transmitted by satellite.[4] Since then, satellites have become increasingly important for the transmission of data, voice, and television signals. Between 1974—when the first domestic communications satellite was launched—and 1984, hundreds of such satellites have been used for communication purposes.[5]

Most communications satellites are located in geostationary orbit, 22,300 miles above the earth's equator. Satellites that are used by U.S. companies for the transmission of domestic television programs are located in this orbit, and as a result, the countries situated between the equator and the continental United States fall within the satellites' "footprint" or the area in which those satellites transmit. Furthermore, some of the geostationary satellites are capable of covering up to 40 percent of the earth's surface with their signal. Thus, with three satellites covering the whole world, it is obvious that nearly everyone will be within the "footprint" of at least one satellite, and will be able to receive its signals, whether or not they are intended for them.

For purposes of copyright analysis, three types of communications satellite systems are distinguished: point-to-point, distribution, and direct broadcast satellites (DBS hereinafter). The point-to-point system is a direct link between two particular earth stations, and the ultimate users receive the transmission by cable or radio. Distribution satellite systems also use conventional cable or radio to retransmit the signals. The difference between these two systems is that transmissions via distribution satellites are intended for more than one receiver or earth station.[6] At the same time that there is wider distribution of the signal, or more intended receivers, there is also the increased potential for interception of the signal by other parties.[7] DBS, on the other hand, is intended for direct transmission of messages to individual receiving sets, without first converting the signal at an earth station.[8]

The main differences between these three satellite systems are the size of the amplification power and the directionality of the satellite. The more powerful the transmission capability of the satellite, the less need there is for a large and costly earth station which can receive low-power signals.[9] Thus, the earth-receiving stations can use smaller antennas or "dishes," depending on what type of satellite transmission is being received. The parabolic dishes no longer need to be meters wide—smaller dishes, a few feet in diameter, are quite capable of intercepting satellite signals and of receiving clearly defined images.

At the same time, reception of the signal by ground stations has been simplified and made less costly. This has also led to an increase in the interception and illegal distribution of the satellite signals. It is now possible for people with the most rudimentary equipment and small dishes to receive television signals if they are within the satellite's footprint. Thus, many individuals and government entities in the region between the equator and the United States have been engaging in "satellite poaching," receiving programs intended primarily for U.S. audiences without paying any of the usual licensing or copyright fees associated with the broadcasting and distribution of television programs.

The poaching of satellite signals raises many questions,

not only the one of compensation for copyright holders. Among the other issues raised is the question of the impact that U.S. programs have in other countries; whether the prior consent of the receiving country is necessary; what legal sanctions are available? These questions are addressed below.

THE IMPACT OF U.S. BROADCASTS BY SATELLITE ON DEVELOPING COUNTRIES

Since the beginning of communication by satellite, there has been a need for the regulation of these activities, taking into account the sovereignty of each country over its communications, and at the same time, recognizing the need for international regulations concerning radio communications. In 1969, the United Nations Committee on the Peaceful Use of Outer Space (COPUOS) established a Working Group on DBS, to consider not only its technical feasibility, but also its legal and political implications.

Since the early debates on DBS, the United States has been a strong advocate of the "free flow of information," whereas other countries have taken the position that prior consent or prior agreements should be reached regarding the direct broadcasting of television programs to other countries. The United States, however, views prior consent as just one form of censorship, and antithetical to its philosophy.[10]

One of the fundamental issues concerning the "free flow" of information as well as direct broadcasting by satellite is that "freedom" means the continued superior position of the United States regarding the flow of information. The United States, which produces and exports the majority of the world's films and television programs, also controls most of the means of distributing these globally.[11]

The imbalance in the world flow of communications led a few years ago to the call for a "New World Information Order,"[12] one which would establish a two-way flow of communication from the developing countries to the industrialized nations. What is perceived as the actual one-way flow, from the developed countries to the less developed ones (LDCs hereinafter) has to be restructured. As one commentator noted, "a nation whose mass media are dominated from the outside is not a nation."[13]

One concern of other countries is the gradual homogenization of the world's cultures, dominated by American values, American advertising, American television and films.[14]

The polemics generated by the New World Information Order are beyond the scope of this paper. It should be noted, however, that the increasing demand for participation in the world communication process has resulted in a growing awareness on the part of all countries of the need to create a more balanced system of information dissemination, of a more equitable distribution or allocation of communications resources, such as the geostationary orbit/spectrum.[15]

And yet, the increased use of satellites for the distribution of television programs (whether by FSS, BSS, or DBS), and the increasing piracy of these signals merely accentuates the disparity in the flow of information and hastens some of the ill effects—the saturation of the world by American television—which COPUOS' members want to avert. It also broadens the gulf between the countries which have developed communications systems and those who aspire to have greater control over their telecommunications infrastructure.

While the United States has over thirty communications satellites in orbit, the LDCs have but a few in operation: two Indian satellites, two in the Indonesian Palapa system, one Brazilian and one Arabsat satellite launched in February 1985. Yet the LDCs account for over 90 percent of the world's population. To date, few developing countries have the capability of distributing their own television programs via satellite—the exceptions being India and Brazil. But it is questionable whether these countries would have a non-domestic market for their productions, let alone a global market. It is very difficult to compete with U.S. technology and the American entertainment industry. Hence, it seems to be less expensive to pirate a satellite signal, and thus obtain an American television program illicitly.

However, this leads to the increased proliferation of U.S.-produced programs,[16] in spite of the opposition which the developing countries have voiced to transmissions without their prior consent. It could be concluded that their prior consent is required only in the case of DBS.[17]

The continued broadcasting from U.S.-based companies, and the continued reception of television by satellite will have a broad impact on the economies of the receiving countries. The entertainment industries, advertising,[18] leisure activities, and employment in these sectors will all be affected, both in the developed countries and the LDCs. Telecommunications affect every sector of the economy, from entertainment and education, to transportation.[19]

In spite of the protests by different nations, claiming that they want to maintain their cultural and national sovereignty, satellite poaching increases. The signals that are most pirated are those intended for U.S. audiences, primarily subscribers to special television broadcast services, such as "Home Box Office" (HBO). The pirating is done by home-installed parabolic dishes, as well as via earth stations that access the INTELSAT space segment. The U.S. companies, which up to now have advocated the free flow of information, are finding themselves in the anomalous position of wanting to be paid for the programs that are being pirated by other countries in the Americas, to compensate copyright owners for the use of their programs. It seems that "free flow" does not mean without pecuniary compensation. It also seems that these smaller nations, which have no highly developed local television or film industries, believe that the benefits gained from pirating satellite signals outweight the costs to their economies—for the time being.

The low cost of parabolic dishes and the proliferation of videocassette recorders,[20] have converted some of the Caribbean countries into the unintended beneficiaries of U.S.-intended TV programs. Hotel owners and other private parties as well as government bodies are engaging in the signal piracy. The poaching occurs in Spanish-speaking countries as well as on English-speaking islands. As the original transmissions are U.S.-made television programs or films, the American copyright owners are the ones who are suffering the most immediate economic harm because of the poaching. But they are not the only injured parties. Local television producers and broadcasters are also affected by signal piracy.

The Motion Picture Association of America (MPAA) alleges that its total revenues from the Caribbean Basin, including television and film sales, are over $20,000,000 a year. It claimed that in Jamaica alone, its revenues from television sales dropped from nearly $100,000 to $60,000 in the course of one year. This decrease in revenues, according to the MPAA, was due to the fact that the Jamaican government had been pirating signals and re-broadcasting them locally, thus bypassing the U.S. TV program producers and suppliers.[21] The MPAA further reported that the gross income in movie theaters on several West Indian islands had fallen because many first-run films, which would normally be shown on the islands a year to eighteen months after their release in the United States, were available to the local television audiences shortly after their U.S. release—via pirated signals.[22]

Although the potential loss of revenue to the MPAA and other copyright owners may be significant, the economic impact of the piracy does not stop there. It will also affect the citizens of the various countries in several ways. For example, if newer films are available on television sooner than at the local theater, it is likely that people will stay at home and watch the TV for a nominal fee, or for free, rather than pay to see an "old" film. This will deprive theater owners of income, but also cause unemployment in certain sectors—theater managers, ushers, projectionists. Furthermore, whatever local television or film production industry that exists will find it very difficult to compete with the multi-million dollar budgets of the U.S. film industry. Obviously, if the local government or a private entrepreneur can provide "quality" programs at very low, if any, cost, there is no incentive to invest in the production of local programs or advertising. This will stifle the emergence of any native industries in these fields. On the other hand, offering U.S.-made television programs to guests at the local hotels may provide an added boon to the tourist industry, a major source of revenue on many of the Caribbean islands.

On another level, the cultural impact of daily American TV fare will be hard to assess, but it is bound to have some influence on the local values and customs of these countries. On the one hand, it might stimulate the desire for local social and economic progress and development. It may also spur some action to protect local folklore and other cultural heritage. On the other hand, it may make the lack of progress more apparent and acute, leading to a sense of deprivation and frustration among the populace.[23] In addition, the differences in languages can become a major problem: many of the Caribbean countries are English-speaking. But there are a number of them where the official language is Spanish or French. Although it is difficult to gauge the impact that American broadcasts will have on the non-English-speaking viewers, undoubtedly it will have some effect. Will it destroy their language, and further contribute to the erosion of their distinctive cultures, or will it act as an incentive to the governments to protect their linguistic heritage as well as their cultural values?[24]

Providing low-cost entertainment to the tourists as well as to the local populations by way of poached satellite signals is saving the governments of these countries a considerable amount of money which can be used for other purposes—to develop other areas of their society or economy. This could be an important inducement to countries already suffering from balance of trade problems, and perennial deficits. Particularly in countries of low per capita incomes, the money saved on developing domestic television programs might be better spent on improving local health care.

The saturation of the broadcasting world with American TV programs and films, and the resulting loss of cultural differences, which has been of concern to the LDCs, is now reality, and aggravated by their own doing. The New World Information Order, which many of the developing countries advocate, will not change the flow of information or of programs from the outside so long as these countries continue to rely on American programs at the expense of their own industries and cultures. These countries are likely to remain passive recipients of information disseminated by a few other countries.[25]

The LDCs need to assess whether the "free" television programs are worth the costs in other areas of their development and intellectual independence. The United States will also have to reassess its position regarding the free flow of information: Should signals intended for American audiences be receivable by anyone possessing an earth dish, or should they be encoded? The United States will also have to consider whether the economic harm to its copyright owners is outweighed by the benefits that other audiences obtain from these programs, or are the costs greater than all the benefits?

With the likely expansion of television broadcasting by satellite, the possibility of signal poaching increases, not only in foreign countries but also domestically. One of the underlying issues, then, is how to protect the intellectual property that is being taken without compensation, how to equitably compensate the copyright owners for the use of their creation, and which— if any—international organization is able to offer and implement this protection?

INTERNATIONAL INTERGOVERNMENTAL ORGANIZATIONS

Several international organisms have been involved since the early days of satellite communications in establishing parameters for these activities. Among these are two specialized agencies of the United Nations, the Committee for the Peaceful Use of Outer Space (COPUOS) and the International Telecommunication Union (ITU). A third global organization, INTELSAT, has provided for the commercialization of a global telecommunications satellite system. Their role, particularly vis-à-vis the piracy of copyrighted television programs relayed by satellite, will be considered below.

The U.N. Committee on the Peaceful Use of Outer Space
Since its establishment in 1959, COPUOS has had the task of formulating norms of conduct for the exploration and peaceful use of outer space. It was noted above that since 1969, COPUOS took under consideration the issues related to DBS, but it has not addressed the problems arising from the use of other satellite systems (FSS or BSS) for television transmissions.

COPUOS formulated some guidelines regarding DBS, and these principles were adopted in 1982.[26] The principles call for the prior consent of receiving countries, but the poaching of satellite transmissions makes the need for prior consent moot.

Regarding the protection of copyright owners, the principles recognize the need to support them, but leaves it up to the different countries to "cooperate on a bilateral and multilateral basis" for the protection of copyright and neighbouring rights.[27]

COPUOS thus leaves the problem of protection of satellite signals and their content to domestic and bilateral agreements. It is not about to formulate an international law that would give global protection to copyrighted transmissions, as this is beyond the ambit of this agency.

The International Telecommunication Union

The International Telecommunication Union (ITU) is a specialized agency of the United Nations which, *inter alia*, formulates technical rules and regulations for the use of the radio frequency spectrum, including those that are to minimize signal interference between the different users. The International Frequency Registration Board, an administrative body which is part of the ITU, provides technical guidelines for the use of radio frequencies used by satellites. The IFRB is provided with basic operational data on any planned satellite system, which is then coordinated with the administrations of other countries. The goal is to design systems in a manner that will minimize satellite transmissions over foreign territory.[28]

As a result of the 1979 World Administrative Radio Conference, developing countries were given preferred access to certain frequencies in the radio spectrum. Under the new regulations, the entire geostationary orbital arc of Region 2 was open to both fixed satellite services and broadcast satellite services.[29]

In the Americas both the FSS and the BSS shared the same frequency band, but after the 1979 WARC, the allocation for the two services was separated. As a result of the ITU's Regional Administrative Conference (RARC) in 1983, the 12.1–12.2 GHz band was allocated to the FSS and the 12.2–12.3 GHz band to the BSS.[30]

The ITU has played a major role in establishing regulations for communications satellites, but these regulations are primarily of a technical nature.[31] Whereas the ITU Convention is legally binding on its members, its regulatory arrangements are not compulsory, nor does the ITU have any explicit enforcement powers. However, because communications by radio—and satellite—require the cooperation of the parties, the regulations are observed by most countries out of self-interest.[32]

The ITU cannot control the content of communications, as it merely recommends the technical parameters for these. The Radio Regulations (Article 17) suggest that each country adopt its own domestic measures to prevent the interception of signals by third parties, especially when these communications are not destined for the general public.[33] However, the ITU leaves the enforcement of this regulation to domestic law. The ITU cannot act as arbiter or censor. Like COPUOS, the members of the ITU believe that protection of the content is not its province. Rather, the ITU leaves it to domestic law to adopt its own measures to protect copyright owners whose transmissions are being pirated, as it has considered that it is beyond the scope of its Convention to regulate the content of any transmission.[34]

The International Telecommunications Satellite Organization

The International Telecommunications Satellite Organization (INTELSAT) is an international, intergovernmental consortium with "the aim of achieving a single global commercial telecommunications satellite system as part of an improved global telecommunications network."[35] Its prime objective is "the provision, on a commercial basis, of the space segment [telecommunications satellites] required for international public telecommunications services of high quality and reliability to be available on a non-discriminatory basis to all areas of the world."[36]

INTELSAT's members and users are the same entities. As signatories to the INTELSAT Agreements, they also establish the general rules concerning, *inter alia,* the approval of earth stations for access to the INTELSAT space segment.[37] Earth station facilities are usually owned and operated by domestic telecommunications entities, and are used primarily for voice and data communications (although some of them are also used to pirate television programs from other U.S. satellites). INTELSAT, however, does provide transmission of television services of a "public interest" nature (current events of global interest) which are not protected by copyright laws. Television transmissions by INTELSAT have expanded, but still account for a small part of the services that this organization provides to its member-users.

As INTELSAT is owned by its member countries, it is unlikely that any of these states would try to impose any control on one another on the use they make of their earth stations that access the INTELSAT space segment. This would be considered an interference with domestic telecommunications policies, contrary to INTELSAT's principles and to general principles of international law. Thus, even though INTELSAT facilities make the poaching of satellite signals possible in some countries, INTELSAT itself is not in a position to censor or even suggest to its members in what activities they should engage. It is not in the position, either, to offer any kind of protection—economic or technical—to prevent any poaching of copyrighted television programs or films intended for U.S. audiences, particularly as the signals intercepted are not from INTELSAT satellites.

It appears that the three major international organizations whose activities relate directly to communications by satellite (COPUOS, the ITU, and INTELSAT) are not in a position, nor should they be put in such a position, to control the use that is made of satellite signals. Whether this use is legitimate or not, or whether the signal comes from FSS, BSS, or a direct broadcast satellite is not relevant. Other organizations are better equipped to deal with the theft of satellite signals, and these will be considered below.

INTERNATIONAL COPYRIGHT CONVENTIONS

Plagiarism—the stealing of someone else's intellectual property and using it as one's own—has been known to mankind for centuries. It is only in recent times, however, that owners of intellectual property have received any protection and compensation for their pursuits.

The development of the printing press in the fifteenth century and consequent proliferation of printing and publishing of books necessitated the protection of the investment of publishers. In England, the Statute of Anne of 1709 granted certain privileges to printers and publishers, enabling them to protect their economic investment. The economic rights of authors were secondary to those of the printers and publishers. In addition, the publishers were granted exclusive rights of reproduction and distribution of the work, the right to protection for a limited number of years, and remedies for infringement of these rights.[38]

In the United States, the rights of authors were incorporated in the Constitution, Article 1, Section 8, which states that authors and inventors will be given the "exclusive Right to their respective Writings and Discoveries [for limited Times]."

Copyright protection generally means that certain uses of a work are lawful only if they are done with the authorization of the owner of the copyright. Among these exclusive rights are the right to authorize broadcasts of the authors' works.

Copyright laws are national in character—they are concerned with acts accomplished or committed in the country itself. The copyright protection is effective only in the country concerned—it is not applicable extraterritorially. Protection in foreign countries is obtained through bilateral agreements and/or international treaties.[39] The two major international copyright conventions are the Berne Convention for the Protection of Literary and Artistic Works, and the Universal Copyright Convention (BU and UCC respectively hereinafter).

There are several other conventions which protect other rights that are not protected by the BU or UCC. The Rome Convention of 1961 offers great protection to broadcasting organizations. The 1974 Brussels Convention Relating to the Distribution of Programme-Carrying Signals Transmitted by Satellite is the first international convention specifically concerned with the protection of satellite signals, but it does not protect copyright owners.

The provisions of these different conventions are applicable only to the signatory countries. As the United States has ratified only the Universal Copyright Convention, it cannot invoke the provisions of the other conventions in seeking protection and/ or compensation from pirating countries. It can only seek remuneration from other countries which are UCC members, or from those with which it has made special bilateral or multilateral agreements.

Both the Berne and Universal Copyright Conventions set forth the protection of authors' rights with respect to broadcasting: Article 11 bis of the Berne Union gives authors of literary and artistic works the exclusive right of authorizing "the broadcasting of their works by any other means of wireless diffusion of signs, sounds or images," and "any communication to the public by wire or rebroadcasting of the broadcast of the work, when this communication is made by an organization other than the original one." However, this article also states that "it shall be up to the legislation of countries in the Union to determine conditions under which the rights in [paragraph 1] may be exercised, but the legislation will apply only in the countries where the conditions have been prescribed."[40]

Article 11 bis (or 11/2) was added to the Berne Convention by the Rome Act of 1928, which consolidated and added two new rights to the rights of authors. Article 11 bis adds the broadcasting right as distinct from the public performance right and also provides for a compulsory license in the Convention.[41]

Similarly, under the Universal Copyright Convention, the authors are ensured the "exclusive right to authorize reproduction by any means, public performance and broadcasting." Article 4 bis of the UCC also provides that any contracting state, by its domestic legislation, may make exceptions to the authors' rights, so long as they "do not conflict with the spirit and provisions of this Convention."[42] These exclusive rights, however, may be preempted by means of compulsory licensing schemes, as provided by national legislation. The compulsory licenses usually provide for rights of remuneration, but not rights of authorization.[43]

Both the BU and the UCC require contractual obligations between author and broadcasting organizations in order to assess the payment of royalties to the author. One drawback to these provisions is that, while they protect only the authors of literary and artistic works, they do not protect broadcasting organizations. Furthermore, even though authors have the exclusive right to authorize the broadcasting of their work, and to equitable compensation therefor, it has been questioned whether these provisions apply to broadcasts by satellite. Both conventions state that the exclusive rights are provided only for those transmissions intended for "direct reception by the general public," thereby precluding satellite broadcasts which are not so intended.[44]

Neither the Berne Union nor the UCC specifically defines "broadcast" or "broadcasting." However, the Rome Convention, which provides protection of neighboring rights and of broadcasting organizations, defines broadcasting as "the transmission by wireless means for public reception of sounds or of images and sounds" [Art. 3, (f)]. Under Article 13 of the Rome Convention, broadcasting organizations enjoy the right to authorize or prohibit certain acts, such as the rebroadcasting of broadcasts, the fixation of broadcasts, the reproduction of such fixations, as well as broadcasts in public places against payment of an entrance fee. Here, too, "it shall be a matter for the domestic law of the State where protection of this right is claimed to determine the conditions under which it may be exercised."[45]

Under the Rome Convention, the exclusive right of authors to authorize or forbid the use of their works disappeared, and the states are obliged only to prevent the distribution of signals by a distributor for whom the signals are not intended.[46]

The applicability of the Rome Convention to satellite broadcasts has been questioned. First, it is doubted whether the protection extends to the originating organization, which converts its programs into signals for relay to the satellite. The receiving earth station, however, which converts the signals and retransmits them to conventional receivers would be protected by this convention's provisions.[47] Second, as the Rome Convention is not universally accepted, and there exists much ambiguity as to its interpretation concerning space circuits, it was discarded as a possible solution to the piracy of satellite signals. The Rome Convention was felt not to apply to pirated signals because "their use would not infringe the broadcasting right recognized by the Convention because the signal is not yet a broadcast in the technical sense used in the Convention."[48]

At the time the Rome Convention was drafted, the question arose whether satellite transmission constitutes "broadcasting." The definition of broadcasting provided in Article 3(f) was interpreted as applying only to transmission by *wireless means* for *public reception* [emphasis added], and reception by an earth station is not considered a "public reception." The narrow interpretation of the Rome Convention excludes satellite broadcasting because the transmission is not for public reception but for reception by an earth station. The wider interpretation is that whatever the technical means employed, the broadcast is eventually destined for "public reception."[49]

The broadcasting organizations point out that they may have no control over who picks up the signal they emit, and, therefore, they cannot be held liable for the communication of a program containing copyright work to an operator and an audience over which they have no control. However, other interpretations view all the phases of broadcasting—the uplink and the downlink—as integral parts of one operation, and therefore within the meaning of the Berne Convention Article 11 bis (I).[50]

Even if a different interpretation were given to the term "broadcasting," so that it would be held applicable to the satellite signal, American broadcasting organizations would not be able to avail themselves of the Rome Convention's protection because the United States is not a signatory to it. Several Caribbean countries are members of this convention, however. Among them Costa Rica, El Salvador, Guatemala, and Mexico. If their respective domestic legislations provided for broadcasting protection, they could claim this protection as against each other, but the United States would not be able to make any such claims.

As a result of the ambiguities in the Rome Convention, it was felt that other solutions were needed for the protection against unauthorized use of satellite-transmitted signals. In 1974, the Brussels Satellite Convention was drafted. This convention is not based on the concepts of copyright or neighboring rights, but rather on the prevention of distribution of signals by distributors for whom the signals are not intended. This convention would create no new rights for broadcasters, but it was hoped that the new treaty would complement the Rome Convention.[51] In the Brussels Convention, "signal" was broadly defined as "an electronically generated carrier capable of transmitting programmes." A " 'programme' is a body of live or recorded material consisting of images, sounds or both, embodied in signals emitted for the purpose of ultimate distribution."[52] The first definition—that of signals—excludes operating signals, and "programme" excludes scientific or technical data, as well as private communications.[53]

The Brussels Convention does not apply where signals are intended for "direct reception from the satellite by the general public." Thus, Article 3 specifically excludes DBS transmissions, because the "originating organization" and the "distributor" are considered to be one and the same under this Treaty.[54] It would apply to FSS and BSS, however.

Another limitation of this convention is that under Article 2, it is left to each contracting State "to undertake adequate measures to prevent the distribution on or from its territory of any programme-carrying signal by any distributor for whom the signal emitted to or passing through the satellite is not intended." The obligation to prevent unauthorized use of the transmissions is on the receiving State, rather than on the transmitting State.[55] The "good faith" of the States in providing effective measures against piracy was to be assumed. The measures to be taken were left entirely to the States' discretion—they could include administrative or penal measures, as well as telecommunications laws or regulations.[56] As the treaty deals with international copyright prevention of signal poaching, other measures could be supplemented by agreements between the different States.

The Brussels Treaty exempts certain contracting States from its provisions: The contracting States are not required to apply "adequate measures" if the unintended distribution occurs in a developing country, and the "programme carried by the emitted signal . . . is solely for the purpose of teaching."[57]

This particular provision could provide an excellent rationale for the Caribbean poachers: for the most part they are "developing countries as defined by the United Nations" and they could claim that their distribution of the pirated signal is "strictly" for teaching and educational purposes. Then the use of the satellite signal would fall under the "fair use" principle, thus not subject to copyright laws. (The Berne Convention and the UCC also make special provisions for developing countries and their use of copyrighted materials).[58]

The provisions in the various conventions making exceptions for the developing countries' use of copyrighted materials for "systematic instructional activities" are subject to these countries obtaining compulsory licenses for the use of these works, thereby ensuring the equitable remuneration of the copyright owners.

The Brussels Convention does not provide protection to copyright owners: one commentator noted that this convention deals with the "container," not the "content," so that the signals are protected, but not the programs emitted by them.[59] The convention does not create any economic rights for authors or other copyright owners, as it is not based on concepts of copyright law.[60]

Even though the Brussels Treaty is not a copyright convention, like the Berne and Universal Copyright Conventions, it is subject to the principle of national treatment: persons protected by the conventions can claim in all contracting States the protection that national laws grant to their own nationals. Nicaragua, Mexico and the U.S.A. are now signatories of the Brussels Satellite Convention; thus, the American signal transmitters are able to seek protection from satellite signal pirates under this treaty in Region 2, but only from the other signatories. Furthermore, this convention protects the signals, not the content, and the claims of the U.S. companies relate to their compensation for the unauthorized use of the programs—the content.

The Brussels Satellite Convention has been ratified by very few countries since it was drafted in 1974.[61] Obviously, its provisions have failed to receive universal support. The United States' interest in this convention has increased recently, especially with the increasing piracy of American satellite signals. Even the U.S. Copyright Office, in a recent report, had recommended its prompt ratification which was effected early in 1985.[62] However, even with ratification of the treaty, the U.S.A. will be unable to receive adequate protection of its interests, unless the countries in which it would be claiming such protection have effective domestic legislation providing sanctions against signal piracy. As in many instances it is the government or State-controlled telecommunications facility which is engaged in the piracy, it is unlikely that it will prosecute itself.

One of the difficulties of copyright owners who seek compensation for the unauthorized use of their works is that each country has varying levels of protection for copyrighted works. Thus, "national treatment" may be minimal or nonexistent, especially regarding broadcasting rights, which can be preempted by compulsory licenses administered by the country in which such license was granted. The remedies an injured party may be able to obtain are those it might contract for individually or by way of bilateral agreements.[63]

At present it would seem that U.S. copyright owners have little recourse by way of copyright laws to seek, and receive, compensation for the pirating of satellite signals in Region 2. Other means of compensation have been sought, with some apparent success.

In 1983, the Caribbean Basin Economic Recovery Act was passed by the United States Congress. Title II of this law conditions the Caribbean States' eligibility for assistance and import tariff benefits on compliance with several criteria. One of these provides that:

> the President shall not designate a country a beneficiary country if a *government* owned entity in such country engages in the broadcast of copyrighted material, including films or television material belonging to the United States copyright owners without their express consent.

Another provision states that:

> the President shall take into account, in making his determination, the extent to which such country prohibits its *nationals* from engaging in the broadcast of copyrighted material, including films or television material belonging to the United States copyright owners without their express consent.[64]

The Caribbean Basin poachers—whether private parties or government organizations—will come under the scrutiny of the United States President in his determining whether their country will be eligible for the special economic assistance that will be provided by this act.

One of the difficulties in relying on international treaties for the provision of remedies is that the treaties are enforceable only as against the governments which are parties to the treaties, and are enforceable through diplomatic channels. The American government may be able to apply certain economic sanctions against the governments of the countries engaged in satellite signal piracy, but there is little, if any, recourse to be had against private parties unless remedies have been provided for on a contractual basis.

However, if there is no contract between the U.S. copyright holders and the poachers, only the threat of economic sanctions will lead to compliance with the CBI's requirements. These may be effective threats, as the following example attests: "[the island of Antigua] was excluded from a list of those eligible for CBI's trade concessions, aid programmes and other economic benefits" because Antigua's national television service, ABS, was making unauthorized use of U.S. satellite television signals by retransmitting them locally. Since the delay in being considered eligible for economic aid under the CBI [Caribbean Basin Initiative] Antigua has made arrangements with several program sources.[65] The U.S. government policy is to withhold economic assistance until there is contractually agreed compliance; in this instance it has proved to be effective.

On the other hand, local governments might hesitate to prosecute private parties who are using pirated programs, particularly if the State-controlled broadcasting organization is also engaged in pirating.[66] Their own "unclean hands" would prevent them from effectively prosecuting the private pirates.

Another point that needs to be considered is the actual economic harm which the U.S. copyright owners are experiencing. The Motion Picture Association alleged that it lost about $40,000 worth of revenues in one country in one year.[67] Compared to its gross revenues of over $20,000,000, this is a paltry sum. Furthermore, the United States is receiving copyright payments from the countries involved in satellite piracy, through the World Intellectual Property Organization (WIPO) which administers the

Berne Convention and other copyright conventions. The American copyright owners also receive compensation through UNESCO, which administers the Universal Copyright Convention, for the use of U.S. television programs and films.[68]

The nonpayment of copyright royalties remains a serious problem, however, and one which is bound to grow.

In a recent report by the U.S. Copyright Office, it was stated that the "copyright industries are losing $1.5 billion in foreign earnings each year through unauthorized use of works copyrighted in the U.S." Another report by CBS notes the U.S. International Trade Commission alleges that the losses run between $6 and $8 billion yearly, due to foreign counterfeiting copyright and patent infringement.[69] These reports, however, fail to state how much income is earned by U.S. copyright holders. Obviously, satellite piracy cannot account for losses in the billions of dollars, but it is a serious economic problem for U.S. copyright owners.

There is also some discontent with foreign countries who impose foreign currency restrictions and other obstacles in the repatriation of profits, and with those countries who impose "cultural restrictions" as trade barriers. The American producers whose products already account for nearly 75 percent of the world market in television programs are complaining of quotas imposed on them by certain countries![70] The focus of these two reports is primarily on the economic and trade aspects of copyright. As the thrust of American copyright law is also economic compensation rather than protection of the moral rights of authors, it is fitting that "cultural restrictions" are regarded as trade barriers rather than as a desire by other countries to promote and protect their own authors and creative industries.

At the same time that the executive branch was seeking to impose economic sanctions on the countries engaged in poaching satellite signals with the Caribbean Basin Initiative law, another branch of the government was authorizing more satellite broadcasts by American companies to the same area. In 1981, the Federal Communications Commission, in its *Transborder Satellite Video Services* order, authorized several U.S. corporations to extend their existing services as well as to provide new television channels via American satellites to several Caribbean islands and countries, among them the Caymans, British West Indies, Bermuda, and Costa Rica. The FCC authorized these services as being in the "national interest" and not contrary to any foreign policy objective of the United States. The latter conclusion was based on its estimation that the new services were only "transborder," and not international, as they are "merely incidental to domestic [U.S.] services."[71]

Because of the potential economic harm that these services would cause to the services provided by INTELSAT in this region, COMSAT argued that the damage would be immeasurable, and also contrary to INTELSAT agreements.[72] However, the FCC apparently dismissed these arguments as unsubstantial, and decided that the television services proposed for the Caribbean countries would cause little foreseeable economic harm. The FCC did not specify which party would suffer "little economic harm," but obviously, the American corporations would not be adversely affected. On the contrary, the transborder extension of these services would provide additional revenues to the U.S. television programmers and carriers,[73] goals which are considered to be in the national (U.S.) interest.

The commission further considered that the 1962 Communications Satellite Act, INTELSAT agreements, and U.S. international telecommunications policies allowed the FCC to authorize the use of domestic facilities for the provision of international (voice and television) public telecommunications services.[74]

The Department of State, in its advisory capacity, cautioned the FCC that even though it might be in the interest of the United States to use domestic satellites for public telecommunications with nearby countries, "certain types of services may be of concern in the minds of neighboring governments. Their concurrence in all instances should not be assumed.[75] At least the State Department was cognizant of the fact that the receiving countries' prior consent, and that INTELSAT's approval, would be required.[76]

Subsequent to the adoption of this FCC order, the National Telecommunication and Information Administration

(NTIA) asked the FCC for a stay of its decision, at least until the United States received assurances that the "interests of the U.S. copyright owners [would] be adequately protected."[77] The United States will have to negotiate with the members of the Berne Union regarding the copyright protection which will be given to the American owners. As the United States is not a member of the Berne Union, whatever protection is sought in Bermuda, the British West Indies, Costa Rica, and the Bahamas (all BU countries) will have to be obtained through bilateral agreements, rather than through the provisions of this copyright convention.

It seems that the different branches of the U.S. government not only have different goals and policies regarding international telecommunications, but that they are unaware of each other's decisions and the impact these varying stances will have abroad. The FCC's position seems to be that what is good for American carriers must be good for their "transborder" clients. The Department of State, officially the foreign policy spokesman of the American government, seems to be aware of some of the implications of the FCC's decision, but does not really spell these out; it merely cautions that the needs of the "transborder" governments should be considered. On the other hand, the NTIA, an agency of the Commerce Department, emphasizes the need to protect the American copyright owners, without taking into account the economic impact the FCC's decision and these different policies will have abroad.

Thus a full circle has been drawn: beginning with the fears of the developing countries regarding their inundation by U.S. television programs, and ending with these services being described as merely incidental to domestic services, and not really of consequence in international telecommunications. The FCC's authorization of these new "transborder" satellite video services will merely add to the continued dominance of American-originated communications services, without necessarily adding to the development of local communications systems in these countries. While the economic gains of American carriers, and the protection of American copyright owners are important goals, they should not overshadow the concerns and needs of the rest of the world.[78]

None of the parties involved in the problem of pirating

satellite signals—the MPAA, the governments abroad, the U.S. carriers and U.S. government—has addressed itself to the long-term economic, social, and cultural impact that the piracy will have on local populations, international telecommunications, and international relations. Although the threat of economic sanctions and retaliation on the part of the U.S. government may act as a deterrent to some countries, it is unlikely that the signal piracy by private parties will end. With DBS on the near horizon and the growing lack of distinction between fixed and broadcast satellite services, these problems will just be accentuated, as copyright infringement will be made easier and less subject to control.

Short-term, as well as long-range solutions to the poaching and other related problems need to be developed.

ALTERNATIVE SOLUTIONS TO THE UNAUTHORIZED RECEPTION/USE OF SATELLITE-TRANSMITTED COPYRIGHTED MATERIAL

To date, the international copyright conventions, such as the Berne Union, the UCC, the Rome and Brussels Conventions, offer few viable solutions to the problem of signal piracy. Except for the Brussels Satellite Convention, the goal of the international copyright treaties is to protect intellectual property and to prevent its unauthorized use. The level of protection differs from country to country, and few copyright laws have provisions dealing with the new technologies.[79] The traditional remedies for copyright infringement—seizure of the infringing works or injunctive relief—are not suitable to satellite signals, and especially not in an international setting. Furthermore, the courts of different countries might be reluctant to grant jurisdiction, let alone substantive rights and remedies to the foreign plaintiff.[80]

Bilateral or multilateral agreements regarding copyright infringements are difficult to enforce, except as between nation-states, which leaves private parties without a forum. These bilateral agreements also involve questions of foreign policy and international relations, and copyright protection is not at the core of foreign policy concerns.[81] (Even in the Caribbean Basin Initiative Act the remuneration of copyright owners is but one of the many factors to be considered in its application.)[82]

One suggested alternative is the establishment of yet another copyright convention which would offer protection in the areas not covered by the Rome or Brussels Satellite Conventions. However, it is recognized that it takes years for international conventions to be drafted and adopted.

Included in the proposal for a new international convention is the establishment of a fee-collecting agency.[83] But this agency might be redundant, as organizations already exist for the administration of the copyright conventions. The World Intellectual Property Organization (WIPO) administers, *inter alia*, the Berne Union and Rome Convention. WIPO and UNESCO, therefore, would be in the best position to establish a blanket licensing system, to collect the licensing fees and disburse them. A special fee could be assessed on the users of satellites for the transmission of television and other entertainment programs. This fee, in turn, would be shared among the owners of copyrighted material that is distributed by satellite.

One benefit of having a supranational licensing arrangement is that all users of satellites for entertainment purposes would be subject to the licensing fee. This would obviate the difficulty encountered with present licensing schemes, which are limited to the country where the license has been established by national legislation, and its application is limited territorially to that country.[84] While compulsory licenses are administered by each country, and the copyright owner's authorization to use his work is not required, the owner is still compensated for the use. A similar licensing arrangement, but global in scope, would fulfill the same function. The amounts assessed from satellite users would be based on considerations similar to the ones on which domestic licensing fees or other royalties are calculated.

A supranational licensing system would offer better protection to copyright owners, particularly if granting the license were made contingent to the acquisition of any satellite-related hardware. Thus, purchasers of earth stations or individual dishes would pay, as part of their purchase price, a fee which would go to the global licensing agency, whether WIPO, UNESCO, or an amalgamation of the two. Governmental and private entities would thus pay a fee (whether called a levy or a copyright fee would not be of crucial importance) at the same time that they obtained their equipment, whether from INTELSAT or from a private corporation.[85]

A blanket licensing system would be applicable to the existing point-to-point and distribution satellite systems as well as to future direct broadcast systems. It has been proposed that a blanket licensing system for DBS should include the originating organization's obligation to pay royalties based on the area in which it intends to broadcast. The royalty collecting organization would be responsible for payment in those areas of unintended reception, but the primary obligation for royalty payments would rest in the originating organization.[86]

The intricacies of negotiating for global blanket licenses for distribution and direct satellite transmissions might be lengthy, as they would necessarily involve several international organizations and the cooperation of all the emitting organizations and the receiving countries. In the meantime, the piracy and other unauthorized use of copyrighted material is likely to continue, and the copyright owners will still not be receiving any compensation.

It is submitted that a more expedient solution to the increased unauthorized interception of satellite signals is to scramble the signals that are transmitted by satellite, whether fixed, distribution, or future direct broadcast systems, particularly if the transmissions are intended for a limited number of subscribers or for particular audiences. Scramblers for both analog and digital signals are available. Furthermore, with the increase in digital transmissions, the encoding would be less cumbersome and less expensive. It would also be more efficient to encode the downlink portion of the signal, as the encryption could be done on the satellite itself.[87]

HBO is reported to be providing scrambling devices free of charge to the 6,200 cable television systems carrying its programs. Additional devices may be obtained for a cost of approximately $1,000 each. The encryption will be centrally programmed and patterns changed when necessary.[88] Other systems available provide for the encryption of only the audio portion of the transmission. The video part can also be encoded, but at greater cost. It is estimated that these cable addressable baseband converters will sell for approximately $150 each.[89]

These two encryption systems are but two examples of the range of possibilities that presently exist. The distributors of satellite-transmitted television programs will have to determine whether the investment in the encryption systems will be worth the cost. Clearly, if HBO is willing to invest close to $10 million in its satellite scrambling system, it is because this will substantially curb its present loss of revenue through signal theft.[90]

It is submitted that encoding satellite signals, together with the establishment of a global blanket licensing system, would drastically reduce the incidence of signal piracy and concomitant loss of remuneration to copyright owners. Both of these solutions are applicable to existing satellite communication systems, including DBS.

Additionally, the developing countries would be in a better position to control the types of programs they would want to receive—they could obtain decoders only for specific signals. Their "prior consent" would be obtained upon their making the necessary arrangements for the decoder and the blanket license. They would be in control of the reception of satellite signals, and could avoid being swamped by unwanted foreign programs by decoding only the wanted transmissions.

On the other hand, the U.S. position on the "free flow" of communications would not necessarily be altered. The U.S. communications companies would be free to transmit whatever they wanted, upon payment of a fee for the use of the satellite signal and the programs. As these would be encoded, their reception would be possible only by willing viewers, intended audiences, those who have obtained the necessary decoders.

With the growth of television broadcasting by satellite, signal piracy and its related problems will become even more pressing issues. A prompt solution—the encryption of programs coupled with a global blanket licensing system—will do much to stem the piracy of signals, to compensate the owners of copyrighted television materials as well as slow down the trend toward a global cultural homogenization based on American values as seen on television or film.

Communications—the sharing of information, knowledge, and culture—is fundamental to mankind; it is the distinguishing feature which has allowed mankind's development.[91] The contributions of authors, creators of literary, artistic, and other cultural works that enhance the quality of our lives, must continue to be acknowledged. The authors or creators of our cultural heritage are entitled to the fruits of their works and should receive compensation for their creations, whether they are communicated to society by way of the printed media or transmitted by satellite signals.

NOTES

1. Several books and reports attest to the need to close the "technological gap" between the developed and the industrializing countries, by placing greater emphasis on the development of their telecommunications infrastructure. In this respect, see: Maitland Commission Report, "The Missing Link: Report of the Independent Commission for World-wide Telecommunication Development," Geneva, December 1984; Report of the Independent Commission on International Development Issues, *North-South: A Programme for Survival* (Cambridge, Mass.: MIT Press, 1982); MacBride Commission Report, *Many Voices, One World: Report by the International Commission for the Study of Communication Problems* (London: Kegan Paul; New York: UNIPUB; Paris: UNESCO, 1980).

2. See *Broadcasting*, April 9, 1984. There are presently more than thirty communications satellites in use by the United States alone, each with approximately twenty-four transponders used for entertainment purposes.

3. A. Clarke, "Extra-Terrestrial Relays: Can Rocket Stations Give World-Wide Coverage?" *Wireless World*, October 1945, pp. 305–308.

4. W. von Braun and F. Ordway, *History of Rocketry and Space Travel* (New York: Crowell, 1975), pp. 201–214.

5. *Broadcasting*, April 19, 1984, pp. 43 ff.

6. The Radio Regulations of the International Telecommunication Union define these different services: (1) Fixed Satellite Service (FSS): a radio communication service between earth stations at specified fixed points when one or more satellites are used; in some cases this service includes satellite to satellite links, which may also be effected in the *intersatellite service;* the fixed satellite service may also include *feeder links* for other space radio communication services (Section III, 3.3). (2) Broadcast Satellite Service (BSS): a radio communication service in which signals transmitted or retransmitted by space stations

194   SYLVIA OSPINA

are intended for direct reception by the general public (Section III, 3.18). (3) The ITU further defines individual reception (in the broadcasting satellite service): reception of emissions for a space station in the broadcasting satellite service by simple domestic installations and in particular those possessing small antennae (paragraph 5.14). (4) Community reception (in the broadcasting satellite service): The reception of emissions for a space station in the broadcast satellite service by receiving equipment, which in some cases may be complex and have antennas larger than those used for *individual* reception, and intended for use: by a group of the general public at one location; or through a distribution system covering a limited area (paragraph 5.15; emphasis in original).

    7. D. Cannon and G. Luecke, *Understanding Communications Systems* (Fort Worth: Texas Instruments, 1980), ch. 10.

    8. Direct Broadcast Satellite systems (DBS) in the United States are still in their infancy, although the FCC authorized several companies a few years ago to build satellites powerful enough to beam television images directly to homes. In November 1983, United Satellite Communications, Inc. (USCI) began transmitting direct broadcasts to subscribers in central Indiana. See *Broadcasting*, April 2, 1984, but this company is "foundering for lack of sufficient capital." See *Broadcasting*, December 12, 1984, pp. 46–54, which states that DBS is a "medium collapsing under its own weight."

    9. P. D. Nesgos, "Canadian Copyright Law and Satellite Transmissions," *Osgoode Hall Law Journal* (1982), 20(2):232–249.

    10. H. Schiller, "The Free Flow of Information—For Whom?" in G. Gerbner, ed., *Mass Media Policies in Changing Cultures* (New York: Wiley, 1977), pp. 105–115.

    11. T. Varis, "The International Flow of Television Programs," *Journal of Communication* (1984), 34(1):143–152, gives the results of a major study regarding the flow of television programs. The United States provides over three-fourths of all television fare for Latin America, whereas in other regions of the world it accounted for 30 to 50 percent of imported programs.

    12. M. Masmoudi, Tunisian ambassador to UNESCO, is credited with coining the phrase "New World Information Order." See MacBride Commission Report, *Many Voices, One World.*

    13. Schiller, "Free Flow of Information," p. 112.

    14. See, for example, T. M. Denton, "Canadian Responses to American DBS Services: Protecting the Infant Industry," paper delivered at the 11th Annual Telecommunications Policy Research Conference, Annapolis, Md., April 24–27, 1983.

    15. The ITU's 1985 World Administrative Radio Conference (WARC-ORB) is charged with the task of "guarantee[ing] in practice for all countries equitable access to the geostationary-satellite orbit and the frequency bands allocated to space services" (Resolution 3 of the 1979 WARC), while "taking into account the special needs of the developing countries and the geographic situation of particular countries." Article 33, ITU Convention, 1982.

    16. Varis, "International Flow of Television Programs."

    17. UNGA Resolution: resolution adopted by the U.N. General Assembly on the Report of the Special Political Committee (A/37/646), 100th Plenary Meeting, December 10, 1982. The "Principles Governing the Use by States of Artificial Earth Satellites for International Direct Television Broadcasting" call for the prior consent of the government of the receiving country. See *Annals of Air and Space Law* (1983), 8:533–538, for full text of the resolution and of the principles. See *Broadcasting*, November 29, 1982, pp. 30, 31, for one view of the effect of the adoption of these principles. For other views on DBS and prior consent, see "International Broadcast Regulation: The North-South De-

bate," *American Society of International Law* (April 1980), 74:298–321; *Syracuse Journal of Interntional Law and Commerce* (Summer 1981), 8:2.

18. Regarding the effects of advertising on the development of the broadcasting media in Latin America, see F. Fejes, "The Growth of Multinational Advertising Agencies in Latin America," *Journal of Communication* (1980), 30(4):36–48.

19. A. Vargas, "A Challenge for the Third World," *Intermedia* (July/September 1982), 10(4/5):29. See "Canadian Responses"; Denton; Maitland Commission Report; MacBride Commission Report.

20. Parabolic dishes range in price from $400 for a device that lets the owner rotate the dish by remote control, to $5,000 for a 12-foot dish. *Business Week,* April 30, 1984, p. 129. For a "global" survey on home video equipment, see *Intermedia* (July/September 1983), 11(4/5):40–64.

21. MPAA (Motion Picture Association of America) Report, "Unauthorized Interception and Transmission of U.S. Domestic Satellite Signals in the Caribbean," Conference on New Developments in International Telecommunications Policy, co-sponsored by the Federal Communications Bar Association and the International Law Institute of Georgetown University, Washington, D.C., May 12–13, 1983.

22. *Ibid.,* p. 247.

23. See MacBride Commission Report and *North-South: A Programme for Survival.*

24. For example, a few years ago the French government undertook to purge from the French language "Anglicisms" that were creeping in, such as "le week-end" (to revert back to "La fin de semaine"), while also eliminating "le businessman," etc. The campaign's success was/is dubious.

25. Varis, "International Flow of Television Programs," p. 152.

26. UNGA Resolution. See n. 17 above.

27. Section H(11) of the Principles Governing DBS; see n. 17 above on copyright and neighbouring rights. Neighbouring rights are those that originally were in the "neighbourhood" of copyright, and related to performances, phonograms, and broadcasting rights of authors. See Rome Convention for the Protection of Performers, Producers of Phonograms and Broadcasting Organizations (1961).

28. M. Rothblatt, "ITU Regulation of Satellite Communication," *Stanford Journal of International Law* (Spring 1982), 18:1–25, at 9.

29. *Ibid.,* p. 13.

30. E. D. DuCharme, M. J. R. Irwin, and R. Zeitoun, "Direct Broadcasting by Satellite: Development of the International Technical and Administrative Regulatory Regime," *Annals of Air and Space Law* (1984), 9:267–288.

31. C. Christol, "The International Telecommunication Union and the International Law of Outer Space," *Proceedings of the Colloquia of the International Institute of Space Law* (1979), 79:39.

32. G. Codding and A. Rutkowski, *The International Telecommunication Union in a Changing World* (Dedham, Mass.: Artech House, 1982).

33. F. Loriot, "Propriété intellectuelle et droit spatial," *Annals of Air and Space Law* (1978), 3:455–462, at 462.

34. J. Lukanik, "Direct Broadcast Satellites," *California Western International Law Journal* (1982), 12:204–230, at 213.

35. INTELSAT, Agreements Between the United States and Other Governments and Operating Agreement. U.S. TIAS 7532, Washington, D.C., August 20, 1971 (entered into force February 12, 1973). Cited as INTELSAT Agreement hereinafter. Preamble. See

N. M. Matte, *Aerospace Law: Telecommunications Satellites* (Toronto: Butterworth, 1982), p. 249 ff., for extracts of the Agreements.

36. INTELSAT Agreement, Article 3.

37. INTELSAT Agreement, Article 8 (b) (v) (A).

38. S. Stewart, *International Copyright and Neighbouring Rights* (London: Butterworths, 1983), ch. 2.

39. WIPO (World Intellectual Property Organization), General Information, Geneva (1983), pp. 40–41.

40. Berne Convention for the Protection of Literary and Artistic Works, Art. 11 bis, s. I (i) and (ii), and Art. 11 bis s. II. Text may be found in Stewart, p. 643.

41. Stewart, *International Copyright*, pp. 94, 95.

42. Universal Copyright Convention as revised in Paris, 1971, Article 4 bis, ss. 1 and 2. Text may be found in Stewart, p. 667.

43. See generally, Stewart, *International Copyright*, pp. 71–76.

44. Lukanik, "Direct Broadcast Satellites," pp. 217–218.

45. Rome Convention, 1961, Art. 13 (a), (b), (c), and (d). For complete text of this convention, see Stewart, p. 679.

46. Loriot, "Propriété intellectuelle et droit spatial," pp. 456–457.

47. Nesgos, "Canadian Copyright Law," p. 237.

48. N. M. Matte, *Aerospace Law*, pp. 202–203, and 200.

49. Stewart, *International Copyright*, pp. 229 and 301.

50. *Ibid.*, pp. 300–301.

51. Report of the Committee of Governmental Experts, Nairobi, 1973, annex I, vol. 53, Internationale Gesellschaft für Urheberrecht E. V., Vienna (1975), p. 193.

52. Brussels Convention Relating to the Distribution of Programme-Carrying Signals Transmitted by Satellite (1974), Article 1(i) and (ii). For text of this convention see Stewart, p. 691.

53. Stewart, *International Copyright*, pp. 252–256.

54. Nesgos, "Canadian Copyright Law," p. 239.

55. *Ibid.*, p. 240; see Stewart, pp. 250–253.

56. Report of the Committee of Governmental Experts, p. 193.

57. Brussels Satellite Convention, Article 4.

58. Berne Convention, Appendix I, Articles 1–4. Universal Copyright Convention, Article 5 ter et seq.

59. Matte, *Aerospace Law*, p. 202.

60. Stewart, *International Copyright*, p. 257.

61. The Brussels Satellite Convention has been ratified by Austria, Federal Republic of Germany, Italy, Kenya, Mexico, Nicaragua, Yugoslavia. WIPO (World Intellectual Property Organization), p. 48. The U.S.A. also ratified it and it came into force on March 7, 1985. Treaty Document 98-3. Information by Department of State Treaty Office.

62. *Broadcasting*, October 8, 1984, p. 81.

63. Stewart, *International Copyright*, pp. 71–76.

64. Public Law 98-67, August 5, 1983, 97 Stat. 384, Caribbean Basin Economic Recovery Act, Section 212 (b) (5) and Section 212 (c) (10), emphasis added.

65. Institute of Caribbean Studies, University of Puerto Rico, *Caribbean Monthly Bulletin*, (November–December 1983), 17(11–12):57.

66. In another instance of poaching satellite signals in Aruba, Netherland Antilles, the High Court of Judicature of the Netherlands held that the 1928 Aruba law regarding radio communication did not prohibit the interception of broadcasts, and there-

fore there was no misuse of satellite "transmissions." February 22, 1985, First Chamber, Petition #6661. The decision of the lower court is reported in *Annals of Air and Space Law* (1984), 9:534–542.

67. MPAA Report, "Unauthorized Interception . . . in the Caribbean," p. 245–249.

68. *United Nations Yearbook* (New York: United Nations Publications).

69. *Broadcasting*, October 8, 1984, pp. 81 and 82.

70. See Varis, "International Flow of Television Programs," pp. 147, 149. "Offered as an example of 'cultural restrictions as a trade barrier' was Canada, which imposes import quotas on programming, which the CBS report alleges has the effect of protecting the Canadian government-subsidized film and television production industry." *Broadcasting*, October 8, 1984, p. 82.

Another report indicates that in Canada, publicly owned networks tend to show more Canadian programming than privately owned ones, but that the United States accounts for the vast majority of imported programming except in the case of the educational network Radio Quebec. This same report states that the United States imports less than 2 percent of its programs. The imports are mostly from the United Kingdom and shown on public television. The other imported programs came from Mexico and Latin America. See Varis, p. 147.

71. *Transborder Satellite Video Services*, 88 F.C.C. 2d, 261–289 (1981) at 268.

72. COMSAT (Communications Satellite Corporation) was created by an Act of Congress in 1962, and it represents the United States in INTELSAT. Article 14 (d) of the INTELSAT Agreements requires consultation with INTELSAT if a party intends to establish or utilize a space segment separate from INTELSAT's to meet its *international public telecommunications* services requirements, to ensure that the proposed system will not cause *significant economic harm* to INTELSAT's global system [emphasis added].

73. 88 F.C.C. 2d at 271.

74. If the FCC's decision regarding transborder services as being merely incidental to domestic operations is upheld, it is plausible that the commission's decision will be to apply wholly domestic legislation to all "transborder" services. Thus, s. 705 of the 1934 Federal Communications Act (as amended in 1984), which prohibits the unauthorized publication or use of communications, might be applied extraterritorially by the FCC or an American court. This section has been interpreted as prohibiting the reception and use of satellite broadcasts by a person unauthorized by the sender to receive the communication. If the unintended recipient happens to be outside the United States, but receiving a "transborder" signal, the FCC just might decide that the pirating would fall under the activity prohibited by s. 705, and within the jurisdiction of the American legal system. The arm of the law can be very long, indeed.

75. 88 F.C.C. 2d, Appendix 1, p. 288.

76. INTELSAT has consented to the provision of transborder services, but only to Canada and Bermuda. *COMSAT* (1984), no. 13. Services to other countries are presumably still under INTELSAT's consideration.

77. *Broadcasting*, September 9, 1983, p. 76.

78. In this respect, it is suggested that the Varis article, "Interntional Flow of Television Programs," be read in conjunction with the comments regarding the CBS report in *Broadcasting*, October 8, 1984, p. 82.

79. One exception may be the U.S. 1976 Copyright Act, 17 U.S.C. ss. 101 et seq.

80. Stewart, *International Copyright*, pp. 47, 130.

81. But see *Broadcasting* magazine's article entitled "Copyright Infringement Tops List of International Problems," October 8, 1984, p. 82.

82. Public Law 98-67, August 5, 1983, 97 Stat. 384, Caribbean Basic Economic Recovery Act.

83. Lukanik, "Direct Broadcast Satellites," p. 223.

84. Stewart, *International Copyright*, pp. 112–113.

85. In France, for example, a tax on the importation and sale of reprographic machines was imposed, with part of the tax being paid to copyright owners of the material copies. In Germany, books in public libraries are subject to a lending fee, to compensate their authors. In addition, other European countries have imposed a levy or tax on the purchase of videocassette recorders, with part of the proceeds going to copyright owners whose works are being recorded at home. This scheme obviates the problem which American copyright owners face in trying to get compensation for the home use/recording of their works. See *Sony Corp. of America v. Universal City Studios, Inc.*, No. 81-1687, 52 USLW 4090 (U.S. January 17, 1984), where the Supreme Court held tht video taping an entire film or television program for "time-shifting" purposes was fair use of copyrighted material, and not an infringement of the copyright. Regarding the European schemes, see Stewart, *International Copyright*, pp. 42 ff.

86. Lukanik, "Direct Broadcast Satellites," pp. 224–225.

87. Information supplied by Gary Schober, WCI Labs, Inc., New York, N.Y., 1984. The Americans are not the only ones concerned with signal interception. It is reported that Arabsat's satellites "will be equipped with on-board decrypters to process the encoded command/control signals sent by ground control stations." *Aviation Week and Space Technology*, May 21, 1984, p. 176.

88. *Broadcasting*, October 10, 1983, pp. 8–9.

89. *Broadcasting*, November 7, 1983, p. 7.

90. *Broadcasting*, October 10, 1983, pp. 8–9.

91. MacBride Commission Report, pp. 3–10.

PART IV: REGULATION OF INTERNATIONAL SATELLITE ACTIVITY

THIRTEEN

A Canadian Perspective on the 1985 ITU Space Conference

WILLIAM H. MONTGOMERY

THE ROLE OF THE ITU

The International Telecommunication Union's 1985 World Administrative Radio Conference will be directed to the planning of the geostationary orbit and the radio spectrum associated with it. To understand the significance of the conference, it is necessary to look at the role of the ITU as an international regulatory body in the context of increased interest by many nations in telecommunications.

The ITU has regulatory and distributive responsibilities in establishing suitable regulations to control the use by its 160 members of the radio frequency spectrum. The regulatory function involves establishing procedures and regulations for coordination of orbital and frequency assignments in order to minimize harmful interference between systems of different countries and facilitate efficient use of the radio frequency spectrum.

The distributive function of the ITU, which has attracted more attention in recent years, focuses on the goal of equitable access to the radio frequency spectrum and the geostationary orbit as well as on the equitable distribution of the benefits from the use of these resources.

International regulatory regimes for the use of the spectrum are established at ITU radio administrative conferences where solutions are sought to conflicting regulatory approaches among member countries. Up to the mid-70s administrations were able to satisfy their anticipated requirements with a fairly rigid approach to frequency planning for certain services. This approach accommodated requirements submitted to particular conferences but did not include modification provisions for the interim period between conferences.

NEED FOR INCREASED FLEXIBILITY

In the 1970s, however, need was determined to establish more flexible means of allocating frequencies. A major turning point in frequency planning was the 1974 Maritime Mobile Services WARC which introduced flexibility into the a priori planning process by adding a procedure for making new entries to the plan as needs evolved.

Another point came at the politically sensitive 1984 High Frequency Broadcasting WARC which adopted planning principles, and a method embodying the major features of plans leading up to this conference. Additional flexibility was introduced in the areas of defining equitable access and satisfying basic requirements, as well as protecting frequency continuity and quality of service.

Technological developments in telecommunications have put a strain on, and encouraged, creative and innovative means of meeting the needs of spectrum users. While technological developments solved certain frequency congestion problems, demand for immediate access to this scarce resource by technologically advanced administrations, and concern by developing countries to ensure access to it when needed, have led to differing views on how to regulate and distribute certain frequencies.

There are many theological arguments for and against

a priori planning, for and against flexible planning, for and against "first-come first-serve," but in our view the question of the space conference is less one of whether to plan than one of how much rigidity is needed in planning the fixed satellite services in order to provide a credible guarantee of access to the orbit and what sort of planning method best serves the domestic, regional, and global interests of the ITU member countries which, of course, reflect the membership of the United Nations.

SPECIFIC ISSUES OF THE CONFERENCE

The 1985 WARC will consider seven planning methods based on submissions from various administrations. These range from the existing regulatory procedures to long-term rigid a priori planning. These alternative approaches to planning methods and their variations will be discussed during the 1985 session.

Rigid a priori planning has the advantage of allowing administrations to implement systems with the certainty that the orbital resource will be available and protected from interference when they need it. The disadvantage is that such rigid plans make it difficult to alter systems at the time of implementation in order to take into account improvements in technology and changes in requirements.

Flexible planning allows for changes at the time of implementation, but the price to be paid is greater uncertainty about the nature of the system that will receive protection from other systems.

Whatever planning approach is accepted, the conference will have to take into account a number of politically sensitive issues. These include developing country concerns regarding whether the spectrum will be available when they require it, without undue and unreasonably costly technological burdens, and satellite operating country concerns to ensure immediate and continuing access to spectrum resources. Procedures will also be needed to incorporate technological change over time and to accommodate the requirements of regional and intergovernmental satellite organizations.

CANADIAN PREPARATIONS FOR THE SPACE WARC

The majority of ITU member nations view the current spectrum regulatory regime as inequitable in the face of a potential shortage of spectrum and orbital resources. The 1985 conference is seen as an opportunity to redress this perceived imbalance in favor of the developed nations, and, conversely, as an opportunity to retain as much flexibility in allocation of orbit resources as possible to ensure continued access to it by technologically advanced nations.

In preparing for the conference, Canada is taking the view that discussion of planning principles cannot be avoided. We have tried to approach the conference in the most pragmatic way we can. As most administrations do, we started with an assessment of our own requirements, of the requirements of the major satellite users, and of what kind of planning principles meet the global requirements for access to the spectrum and permit everyone's space requirements to be met in the future.

The proposals that we are putting forward have been discussed with the U.S. administration, with the British administration, the administrations of the European countries, and with Latin American administrations. The objective of this process is to insure that the ITU Space Conference does not result in a major confrontation between what are essentially developed countries on one hand and developing on the other. It is most essential that we maintain a sense of cooperation and the working relationships in the ITU which are important to all of our administrations.

A working document, presented below, was drawn up with these concerns in mind and has been discussed bilaterally and regionally in order to solicit views and comments of interested administrations. Our objective in putting forward this document was to help lay the groundwork for development of workable proposals for the conference.

The ideas presented in the Canadian working document, encompassing the major points we believe need to be addressed at the 1985 conference should be seen in terms of their representational character. These ideas are based on discussions with various administrations and reflect the breadth of interests of the major users of satellite systems. They are meant to satisfy the objectives of most ITU member countries in providing reasonable assurances of access to the geostationary orbit.

This document does not in any way complete Canada's preparations for the conference. Additional studies are being completed and consultations with other countries will continue right up until the Space WARC. Whatever planning approach is eventually adopted, regulatory procedures are likely to further emphasize the ITU's role as a "guarantor" of access to the spectrum. How this role will be defined will depend on the major features of the approach to frequency planning and associated regulatory procedures that will be included in the outcome of the 1985 conference.

WORKING DOCUMENT: CHARACTERISTICS OF AN A-PRIORI PLANNING METHOD FOR THE FIXED-SATELLITE SERVICE

I. Preamble

This a priori planning method is intended for application to the fixed-satellite service in the heavily-used 6/4 GHz and 14/11-12 GHz bands. This planning method is seen to be a priori and flexible at the same time because it strikes the best balance between seemingly contradictory requirements of providing long-term guaranteed access by administrations (includes "group of administrations") to the use of the geostationary orbit, while allowing sufficient flexibility to introduce cost-effective state-of-the-art systems.

This planning method responds to the need of administrations to have enough spectrum/orbit resource reserved in a precise way for their use when required and at the same time allows management of that spectrum/orbit resource to be responsive to the latest technical innovations available. It does this by specifying a minimum number of parameters, i.e., orbit position, frequency band, and service area, and allowing all other parameters such as antenna characteristics, interference limits, modulation characteristics, spacecraft station-keeping, and pointing-error characteristics of the satellite networks to be determined at the time that those networks are implemented. This results in the maximum possible flexibility within the context of an a priori plan.

Under this method an a priori plan will be developed at the 1988 Space WARC, which will assign specific orbital positions and frequency bands to each administration. As a result, the method allows implementation of state-of-the-art networks to be coordinated under the plan, and avoids the need to specify the detailed characteristics of these systems in 1988 long before they are designed. This balance, or one very close to it, will have to be adopted by WARC-ORB if there is to be an orderly and effective use of the geostationary orbit by the fixed-satellite service in the 1990s. This is consistent with the objectives of Resolution 3 of the ITU Radio Regulations.

II. Characteristics of the Plan

1. Development of the Plan

The formulation of the plan will be based on requirements requested by administrations. These stated requirements will include the frequency band width, the service area, and the service orbital arc of each required network. Separate sets of requirements would be submitted and accommodated for the 6/4 GHz band and the 14/11-12 GHz band, with account taken of the need to meet certain requirements with multi-band hybrid satellites. The number of orbital positions assigned to an administration in the plan would be determined by the conference and would be based on the requirements submitted by that administration.

To develop the plan, supplementary sets (perhaps four) of detailed technical parameters representative of the types of systems which are expected to be implemented during the lifetime of the plan would be developed by the conference to enable the creation of the plan. Each set would correspond to agreed typical satellite systems and include such parameters as antenna gains, interference limits, signal characteristics, spacecraft pointing accuracies, and station-keeping tolerances. Requirements submitted by administrations to the second session would be defined in terms of the applicable set and would be used to develop the entries in the plan.

Once the plan is established these detailed sets of parameters would have very limited use; they would in no way constrain the characteristics of systems when they are implemented under the plan, nor when they are coordinated. They would, however, be used to determine the acceptability of proposed modifications to entries in the plan, i.e., the frequency band, orbital position, or service area.

2. *Entries in the Plan*

Administrations will be given entries in this plan in response to their stated requirements. Each such entry will specify the nominal orbital position, the frequency band or sub-band, and the service area of the network to be implemented under the plan. Unlike the earlier broadcasting satellite plans, no other characteristics than the above would be specified in the plan; they would be left intentionally to be agreed upon at the time that systems under the plan are coordinated.

3. *Accommodation of Existing Systems*

To allow a smooth implementation of the plan, satellites of an administration in orbit at the time that the plan is adopted but not at an orbit position assigned to that administration will be accommodated for the remainder of their notified operational lifetime, at their existing orbital position if possible, or, if not possible, then within their service arc. During this transition period of perhaps several years, existing systems may have to change orbital position within their service arc to accommodate new systems as they are implemented in accordance with the plan. It is understood that the number of changes so imposed on operational systems would be kept at a minimum.

4. *Coordination of Systems within the Plan*

As systems are implemented in accordance with the plan, they will be coordinated based on procedures to be developed at the conference, procedures which will be similar to Articles 11 and 13 and Appendix 29 of the current Radio Regulations. Two situations apply:

a) Under normal circumstances, new systems implemented in accordance with the plan will be coordinated through bilateral or perhaps multilateral discussions between administrations using the actual technical parameters of operational systems and of other new systems that are being coordinated, subject to the orbital positions, frequency bands, and service areas specified in the plan. Coordination will be based upon the latest CCIR recommendations or technical criteria agreed to by the parties concerned. For the majority of cases, it is expected that the systems to be implemented will fall within the set of parameters identified with the original requirement.

b) In exceptional circumstances, where coordination cannot be readily completed, the IFRB will convene a special meeting of the administrations involved in the original coordination to find a means by which the proposed system shall be shared both by the administrations of existing networks and the administration wishing to establish the new network.

5. *Plan Modification Procedure*

A modification procedure would be part of the administrative procedure associated with the plan, to be used where required to make additions, deletions, or changes to the basic parameters of the entries in the plan, i.e., to the frequency bands or sub-bands, orbit positions, or service areas. Modifications would be subject to agreement of affected administrations. The examination of such proposed modifications would take into account the actual characteristics of systems that are already operational or coordinated, and the characteristics originally assumed in the plan for all other entries.

The modification procedure would also be used whenever a change to the basic parameters of the plan was required as a result of the coordination process.

6. *Duration of the Plan*

The resulting plan would be in effect for a minimum of ten years, and would be subject to review after that time. It would remain in effect until such time as it is replaced by a new agreed plan. The ten-year duration is the expected interval over which its technical basis would reflect the technology used to implement actual systems.

FOURTEEN

Access to Information Resources: The Developmental Context of the Space WARC

HEATHER E. HUDSON

A t the 1979 World Administrative Radio Conference (WARC), delegates unanimously passed Resolution 3, which states that a World Administrative Radio Conference is to be convened "to guarantee in practice for all countries equitable access to the geostationary satellite orbit and the frequency bands allocated to space services."[1]

The principles of access and sharing of the geostationary orbit which were first put forward at the Extraordinary Administrative Radio Conference (EARC) in 1963 were further refined at the 1982 ITU Plenipotentiary, in paragraph 154 of the Nairobi Convention:

In using frequency bands for space radio services, Members shall bear in mind that radio frequencies and the geostationary satellite orbit are limited natural resources and that they must be used efficiently and economically, in conformity with the provisions of the Radio Regulations, so that countries or groups of countries may have equitable access to both, taking into account the special needs of the developing countries and the geographical situation of particular countries.[2]

The first session of the Space WARC will be convened in August 1985. In the United States, representatives of government and industry are currently studying options and developing positions concerning the "principles, technical parameters, and criteria for planning" which are to be the subject of this first session.[3] (ITU Administrative Council, 1983).

Yet it is the premise of this paper that the actual agenda of the Space WARC and other international telecommunications negotiations is in fact much greater: it concerns equitable access to the tools of the information revolution—the means of accessing, transmitting, and sharing information that are the keys to social and economic development.

GREATER INFORMATION GAPS?

What progress has there been since 1979? Despite the rapid advances in technology in the past five years, there is evidence that most developing countries are not catching up to the industrialized world in access to these information tools. As of mid-1984, there were 149 commercial satellites in orbit, of which 121 or 81 percent had been launched since 1979. Only fourteen satellites serving developing countries had been launched during this period. Even taking into consideration the fact that INTELSAT satellites are used by both industrialized and developing nations, satellites for the exclusive use of industrialized countries make up 72 percent of the total.

Middle-income countries have benefited most from increased access to satellites for domestic and regional communications. By the end of 1985, twenty-six middle-income countries will participate in domestic or regional satellite systems, or will lease domestic capacity from INTELSAT. Only seven of the thirty-four countries classified as low income by the World Bank will be participants in such systems (see table 14.1).[4]

TABLE 14.1. Status of Domestic/Regional Satellite Services
for Developing Countries in 1985

	Participating Countries	
	Low Income	Middle Income
Domestic satellites	1	2
Regional satellites	5	18
INTELSAT domestic leases	3[a]	15[b]
Percentage of countries	21%	41%
Percentage of population	52%	72%
(excluding India and China)	15%	

SOURCES: COMSAT Satellite Chart, 1984; INTELSAT Fact Sheet, 1984.

[a]2 of these countries are also participants in domestic or regional systems.

[b]9 of these countries are also participants in domestic or regional systems.

Eighteen developing countries currently lease capacity from INTELSAT for domestic use. Of these, seven are oil exporters, and only three are among the least developed countries. It must also be recognized that domestic use of INTELSAT is generally limited to improving communications to provincial capitals or regional centers because the satellite is not designed for low-cost, thin-route telecommunications or community broadcast reception.

There has been very little progress in extending access to telecommunications services within developing countries. Although there are now more than 600 million telephones in the world, it is estimated that two-thirds of the world's population has no access to telephone services. Tokyo alone has more telephones than the whole African continent. Nearly three quarters of the world's population live in countries with ten telephones or fewer for every 100 people; over half the world's population lives in countries with less than one telephone per hundred people, and most of these telephones are located in urban areas. As a result, the telephone density is likely to be much lower in rural areas where telecommunication is more critical because of the difficulty of communicating over longer distances.

Again, lower income countries show dramatically limited access to telecommunications (see table 14.2). The highest telephone density among these countries is seven telephones per thousand people, while in many cases there is no more than one

TABLE 14.2. Telephone Densities: Low- and Middle-Income Countries
for Which Data Are Available

Low-Income Countries	Telephones per 100 Population	GNP/Capita
Bangladesh	0.1	140
Ethiopia	0.3	140
Mali	0.1	190
Burundi	0.1	230
India	0.5	260
Tanzania	0.6	200
Sri Lanka	0.7	300
Pakistan	0.5	350
Mozambique	0.4	350
Ghana	0.6	400
Middle-Income Countries		
Kenya	1.3	420
Indonesia	0.4	530
Bolivia	2.5	600
Honduras	0.9	600
Zambia	0.5	600
Egypt	1.2	650
El Salvador	1.8	650
Thailand	1.1	770
Philippines	1.2	790
Papua New Guinea	1.6	840
Morocco	1.1	860
Nicaragua	1.8	860
Nigeria	0.7	870
Congo, Peoples Republic	0.6	1110
Guatemala	1.4	1140
Peru	0.7	1170
Ecuador	3.2	1180
Jamaica	6.2	1180
Dominica	3.0	1260
Columbia	6.3	1380
Costa Rica	10.9	1430
Turkey	4.7	1540
Syria	4.9	1570
Korea, Republic of	13.8	1700
Iran	2.9	1700
Malaysia	6.3	1840
Panama	10.7	1910
Algeria	3.3	2140
Brazil	7.2	2220
Mexico	7.4	2250
Argentina	10.7	2560
Chile	5.2	2560
South Africa	13.1	2770
Uruguay	10.1	2820
Venezuela	9.4	4220
Hong Kong	35.0	5100
Israel	32.1	5160
Singapore	31.6	5240

SOURCES: World Bank, *World Development Report* (Washington, D.C.: World Bank, 1983);
AT&T, *The World's Telephones* (Morris Plains, N.J., 1983).

telephone per thousand inhabitants. Yet in most middle-income countries, including those with access to satellites for domestic communications, the infrastructure is still extremely limited. Brazil and Mexico have an average of seven telephones per 100 population; again rural densities are much lower. Many middle-income countries have even lower telephone densities, including members of Arabsat and users of the Palapa system.

The Maitland Commission estimates that $8 billion from all sources was invested in telecommunications in developing countries in 1983 and that at least an additional $4 billion per year will be necessary if minimal worldwide access to telecommunications is to be achieved.[5] Efforts to improve utilization of telecommunications through support for developmental applications and for training have also been disappointingly limited. The International Programme for Development Communication (IPDC) was established in 1980. So far, it has generated a long list of proposed projects, but very limited funding. While training has been identified as a priority, again, advances in training opportunities have been very modest. Recent bilateral initiatives include the U.S. Telecommunications Training Institute, established in 1982 in time to be announced at the Nairobi Plenipotentiary. It provides short-term technical training through U.S. industries. However, there has been little progress to expand longer term technical training in countries and to provide training opportunities in communications applications and coordination with user organizations and their representatives, such as educators, health care providers, and agricultural extension officers.

A recent venture by INTELSAT illustrates the possibilities but also the difficulties of planning developmental services. Project SHARE offers developing countries free time on INTELSAT satellites for tests and demonstrations of satellite technology for education and other developmental activities during a sixteen-month period beginning in January 1985. While free time appears to be a significant contribution, it is a very minor component of a development communications project. Planning at the national level must involve the telecommunications officials in collaboration with users from other agencies or ministries such as health and education. Funding must be found to support the project, including personnel, operations, and equipment. Funding is likely to be sought from bilateral or multilateral aid agencies which have their own priorities and funding cycles. Once funding is obtained, equipment must be procured and installed, and project planning activities must begin. Accomplishing all of this within sixteen months or even two years is highly unlikely. However, demonstrations and pilot projects can be combined with longer term strategies to develop a technology appropriate for Third World use. For example, INTELSAT's new VISTA terminals for low density telephone service for rural and remote areas and its INTELNET service for low cost point-to-multipoint service for very small earth stations are examples of steps in the right direction. But financing is still a problem, particularly for low-income countries. And developing countries will be reluctant to invest the time and effort to plan and implement pilot projects without a very strong possibility of their continuation and expansion.

BASIC REQUIREMENTS

As noted above, developing countries require major investments in their infrastructure to meet minimum telecommunications access goals. These goals may be stated in terms of both distance and density, for example, a minimum of one telephone per hundred people in urban areas; or one telephone within an hour's walk in decentralized rural areas; or one telephone per permanent settlement in remote areas.

Solutions to developing countries' basic telecommunications problems are now within reach. However, a variety of initiatives will be needed. Satellites offer a very promising solution to the rural and remote area problem. With systems designed for thin-route service, earth stations of 4.5 meters or less can be used for rural telephony and broadcast reception. The stations can be installed where needed throughout developing regions, without waiting for the extension of terrestrial networks from cities.

Most developing countries do not need and cannot afford their own domestic satellite systems. Regional systems owned by a consortium of countries are promising if the countries can agree upon an organizational structure and obtain the necessary financing. Yet for many developing regions, notably sub-Saharan Africa, Latin America, the Caribbean, and the South Pacific, appropriately designed satellite capacity is not available. For these countries, the most promising solution may be shared capacity on regional or international systems.

The second step is the acquisition of earth stations and other equipment required to link users to the satellite. Developments in technology in the past five years have resulted in smaller and cheaper equipment now in use in the industrialized world that is ready for use in the developing world, e.g., small earth stations with antennas that can be assembled and pointed with only minimal technical skills and can be broken down for shipping in small aircraft or by truck; spread spectrum techniques that allow transmission to very small terminals (e.g., .7 meters to 1.5 meters); solid state exchanges of 100 lines or less; low-cost VHF and UHF links from earth stations to surrounding communities; alternative power sources for areas without electrification.

Many of these technologies have been developed for specific uses within industrialized countries—for example, data transmission from remote oil rigs to central computers, for voice and data communications with oil exploration crews, for telephone and broadcast services to isolated settlements, for dedicated networks for transmission of market information, news and financial services, weather data, etc. Yet these technologies could also serve developing countries' needs as well. For this to occur, developing country planners need to be aware of these new technologies, and mutually attractive arrangements for suppliers and purchasers need to be negotiated. The latter would include low-cost financing coupled with guarantees of prompt repayment in hard currency; training in installation, operation, and maintenance of equipment; and transfer of technology to the extent feasible, ranging from in-country assembly of components to manufacture under license and joint ventures.

The third step involves utilization. The full verdict on satellite applications is not yet in, but it is safe to say that the technology has not been utilized to its full potential for socioeconomic development. The reasons are many, and have been addressed elsewhere.[6] They include lack of coordination between the telecommunications administration and potential users including social service and commerical entities, lack of resources to develop and test applications, and disincentives to innovation within the potential user organizations.

PROPOSED SOLUTIONS: AN AGENDA FOR ACTION

Appropriate Satellite Capacity

A promising near-term solution would be modification of the INTELSAT VI series now under construction, with launches planned to begin in 1986. It appears that a package of transponders with higher powered regional beams could be added to the payload at minimum incremental cost, and within the launch weight limitations. These satellites will be located over the Atlantic and Indian Oceans where they could direct beams for coverage of Latin America and Africa.

The advantage to developing countries would be substantial in that the incremental cost would be far lower than a dedicated national or regional satellite, and each country could negotiate autonomously with INTELSAT for capacity as required, rather than having to reach agreements with individual countries for access to their satellites, or having to create new regional institutions.

The question of financing remains. The cost of leasing or even buying capacity outright for domestic use should be modest enough to be within reach of middle-income countries using available sources of financing. For the least developed countries, financing at concessional rates through the International Development Association and the various regional development banks should be available.

On the Indian Ocean INTELSAT VI, capacity for an African regional network could be added, as well as transponders for unserved Asian countries such as Pakistan or Burma, if required. It is assumed that China will have its own domestic satellite system by the end of the decade. However, interim service for China could also be provided by adding capacity to the Indian Ocean INTELSAT VI satellites.

Similarly, a regional package could be added to an Atlantic Ocean INTELSAT VI for service to Latin America including subregional beams for the Andean nations, Central America, and the Caribbean. Thus domestic and regional service could be provided to all countries not served by the Mexican and Brazilian satellites.

For the South Pacific, the requirement is not for additional capacity, but for a beam with higher power and appropriate gain setting to cover the Pacific Island nations of the South and Southwest Pacific. The requirements for modification are analogous to Alaska's requirements in the early 1970s. As a result of negotiations with the state of Alaska, which required capacity for telephone service for Alaskan villages, RCA modified the transfer gain settings on its Alaskan transponders, thereby making possible the use of small (4.5 meter) antenna earth stations now in use in more than 100 villages. The INTELSAT satellites in the Pacific carry sufficiently low traffic volumes that dedication of a few transponders (say up to four) for intraregional Pacific use could be accommodated. INTELSAT Vs have the capacity for 12,000 simultaneous telephone circuits and two television channels. In 1973, only 3,629 full-time circuits were in use in the Pacific region.[7]

Ground Segment

As noted above, innovative approaches must also be found to ensure that developing countries gain access to affordable and appropriately designed terrestrial facilities. These requirements, which are designed to reduce the risk to suppliers of operating in the developing world and to reduce the risk to developing countries of dependency on external expertise and equipment, include: dissemination of information about available technology appropriate for developing country conditions; financing arrangements that reduce risk to suppliers; training that enables developing countries to take full responsibility for installation, operation, and maintenance; and strategies for technology transfer, varying with the needs and aspirations of countries at various stages of development.

One strategy the United States and other industrialized countries could pursue to meet these requirements, which would in turn open up major export markets for their technology, would be an exhibit in conjunction with the 1985 Space WARC of technology appropriate for developing country use. This would not be on the same scale as TELECOM 83, which emphasized technologies aimed at large-scale users in industrialized countries, but would include small earth stations, transportable uplinks, small exchanges, low power radio and TV transmitters, and solar and wind power supplies. It could be sponsored by the ITU and underwritten by industrialized countries and companies. Another would be a meeting of major lenders for telecommunications including the World Bank, the regional development banks, the Arab fund, the EEC, and private and bilateral lenders to put together a funding package for countries interested in leasing or buying INTELSAT space segment, and obtaining ground segment facilities. The agenda would include both financing and risk reduction strategies such as loan insurance or payment guarantees.

Developmental Applications

In order to encourage applications of satellite technology for social and economic development, including education for children and adults, health care delivery and other social services, support for economic development activities including agricultural extension, cooperatives, and private sector production and marketing, a variety of approaches could be pursued, including:

- creation of a Center for Telecommunications Development within the ITU, as advocated by the Maitland Commission, with staff that could assist not only with technical planning, but also with economic feasibility and evaluation studies;
- support for pilot projects in satellite applications that could be undertaken with existing INTELSAT and regional facilities and with augmented INTELSAT capacity as proposed above. This approach could include an expansion of project SHARE, with support from industrialized countries in terms of donations or loans of equipment and resource people, and from

multilateral and bilateral aid agencies in the form of funds for project development, staffing, and evaluation;

- internships for developing country technical and applications planners and practitioners in agencies and organizations with experience or responsibility for developmental applications of satellite services. These would include: Departments of Communications— e.g., in Canada and Australia; and applications projects such as the Learn/Alaska Network in Alaska, the Learning Channel, the University of Wisconsin Extension Network, the National Technological University in the United States; the Inuit Broadcasting Corporation, the Canadian Broadcasting Corporation Northern Service, Knowledge Network, and Wawatay Radio Network in Canada; the Open University in the United Kingdom; the School of the Air and aboriginal media projects in Australia; and regional Saami (Lapp) broadcasting in Scandinavia.

Other exchanges could take place with developing countries, such as projects sponsored by the U.S. Rural Satellite Program in Peru, Indonesia, and the West Indies; INSAT rural services in India, and the University of the South Pacific in Fiji.

Funding for these internships could be provided by host countries, or could take the form of exchanges between countries, so that both would benefit from the experience. Multilateral funding, for example, from UNESCO and regional broadcasting and telecommunications organizations could also be provided for exchanges between developing countries.

RELATIONSHIP TO THE SPACE WARC

The steps outlined above could make major strides toward achieving the goals of equitable access to information resources for developing countries. If commitments could be made before the first session of the Space WARC in August 1985, they could demonstrate a pledge by the industrialized world to ensure that developing countries are able to gain access to satellite technology. At the same time, they would demonstrate an efficient use of orbit and spectrum resources.

These approaches need to be considered within a framework of policy guidelines based on the principles of equity and flexibility such as:

- accommodation of international and regional systems: although the ITU functions as a body of national administrations, planning for satellite systems needs to include international and regional systems, so that entities are protected that can efficiently serve both industrialized and developing countries;
- flexibility in proposing technical solutions: It must be recognized that many trade-offs are involved in evaluating solutions to orbit utilization. Developing country planners are likely to be wary of technical solutions including computerized models and frequency reuse techniques with which they have little experience or which may result in more costly solutions in terms of hardware cost or complexity;
- assessment of cost: While cost to industrialized countries may be seen primarily in modifications to technical designs or delays necessitated by uncertainty, cost to developing countries may include the number of staff needed to monitor short time-frame plans and procedures, the cost in time and travel for countries to meet frequently to resolve problems or modify the planning process; the perceived danger of depending on imported, overly complex and/or costly technology;
- allocation of costs: To ensure equity or fairness to all entrants requiring satellite orbit locations and spectrum, guidelines must ensure that all participants share the cost of accommodating new systems; and
- verification of requirements: To avoid "requirements inflation" and wasteful squatting on unused orbital

locations, criteria must be adopted to assure that requested capacity is in fact utilized within a reasonable time period; otherwise, they would revert to the pool of available locations and frequencies.

These guidelines should be acceptable to both industrialized and developing countries in that they are designed to promote both equity in access to the geostationary orbit and flexibility to accommodate changing technology, and regional and international requirements. Yet they do not solve the problem of increasing access to the telecommunications tools which developing countries need to acquire, transmit, and share information needed for social and economic development.

However, these guidelines combined with the steps outlined above can address the issues underlying the Space WARC. Failure to view the Space WARC in a developmental context could result not only in an impasse at the conference, but in the perpetuation of inequitable access to information resources that will impede development and prolong dependency.

NOTES

1. ITU (International Telecommunications Union), *Final Acts of the 1979 World Administrative Radio Conference* (Geneva: ITU, 1979).

2. ITU, *Final Acts of the Plenipotentiary Conference* (Nairobi: ITU, 1982).

3. The 1985 Space WARC agenda is contained in Resolution No. 895 which was adopted by the ITU Administrative Council in 1983. Dr. Hudson's paper was written before the conference, as were the other papers in this part of the book.

4. INTELSAT, "INTELSAT Fact Sheet" (Washington, D.C.: GPO, June 30, 1984).

5. Maitland Commission, "Draft Report," London, 1984.

6. Heather E. Hudson, "Satellite Communication and Development: A Reassessment." Paper presented at the Annual Conference of the International Communications Association, Dallas, May 1983.

7. INTELSAT, *Annual Report* (Washington, D.C.: GPO, 1983).

F I F T E E N

The Role of International Satellite Networks

WILSON DIZARD

S atellite networks have transformed world communications in less than twenty years. The changes have been so swift that many of the conditions which guided the first generation of satellite development are outdated, one result of a stunning success in global cooperation.

A major attempt to redefine international rules for satellite development will take place next year. The occasion will be an ITU World Administrative Radio Conference—the so-called Space WARC. The conference will be held in two sessions, the first in August 1985, and the second in 1988. Both sessions will deal primarily with technical and administrative matters. The Space WARC agenda focuses on a review of the procedures whereby the ITU administers access to two natural resources needed for satellite communications. They are radio frequencies and the geostationary orbit (GSO)—the vast circle above the equator where most satellites are placed.

Behind the conference's technical discussions, however, are important political and economic implications, affecting international communications development generally and American interests specifically.

This paper will discuss U.S. strategy for the conference as it relates to one critical aspect of the meeting. This is the paradox: the conference agenda does not deal with the most important players in global satellite operations. These are the multilateral organizations which run the international networks. The most important of the organizations is INTELSAT, the 110-nation consortium which provides services, directly and indirectly, to 175 countries and other jurisdictions worldwide. There are other networks: the Soviets, Europeans, Arabs, and Indonesians have, now or in the near future, smaller systems. Collectively these multilateral organizations are responsible for over 90 percent of all satellite traffic. (The remainder involves primarily U.S. domestic networks.) The multilateral organizations (and, preeminently, INTELSAT) are, in short, the key players in global satellite communications.

INTELSAT and the other networks will be at the Space WARC as nonvoting observers. Their interests will be represented fractionally by their members, who make up 75 percent of ITU's constituency. The reason for this is that the union is, in the UN pattern, an organization of sovereign nations.

As a result, the Space WARC agenda is shaped in terms of national interests. Specifically, its discussions will center around differing views of sovereignty rights as they relate to access to satellite radio frequency and geostationary orbit resources. Is access to these resources essentially a free right of any nation, based on needs and capabilities to use them? Or are they (in the Third World phrase) "the common heritage of mankind," to be allocated equitably to each country on a predetermined formula?

The United States and other "Northern" countries support the former approach. Current ITU procedures generally conform to this relatively unencumbered access, subject to technical coordination standards. As a result, there has been considerable flexibility in the availability of these resources—an important element in encouraging the rapid expansion of satellite networks over the past twenty years. The basic American position going in to the Space WARC will be to preserve this flexibility.

The less developed countries (LDCs) are the ITU majority. They will come to the conference supporting major changes in the present system. Their goal is a regulatory regime which will give them (in two key words) "equitable" and "guaranteed" access to satellite frequencies in certain services and to GSO resources. In most of their proposals, this will translate to some form of exclusive "ownership" of these resources, country by country. Any formula that is adopted will recognize the need to adjust this vesting to such factors as geographical size of an individual country and/or to its population. However, the overall result will be to lock in frequency and GSO resources to a large number of LDCs (e.g., Belize, Nepal, The Gambia) which have no foreseeable plans for developing a national satellite system. The current satellite ground rules will be changed in a significant way. There is a rough analogy to the more complex Law-of-the-Sea negotiations of the past decade.

The stage appears set for another North-South confrontation on a global resources issue. This is, however, too simple a scenario for the Space WARC. As suggested above, it ignores the fact that, day by day, almost all satellite communications are carried out by multilateral organizations, and particularly by INTELSAT. The conference will make its decisions, under current arrangements, on the basis of national claims to access to frequency and GSO resources. Whatever the final decisions, the needs of the multilateral organizations will be squeezed into a national sovereignty formula, as they are now.

This discussion paper will review the implications of this for U.S. strategy at Space WARC. It will examine whether, and how, INTELSAT and other multilateral organizations might play a more active role in proposals for moderating the competing approaches to satellite-resource access which will be submitted to the conference.

The rationale for looking at this prospect is clear cut. INTELSAT is, in reality, the guarantor of equitable access to satellite services for most ITU members, particularly in the developing world. Vesting claims to frequency and GSO resources will have no practical effect on strengthening the opportunities for equitable, guaranteed service for these countries. Such vesting presupposes that a country will develop its own national satellite system—an assumption that does not apply, for economic and other reasons, to most of the ITU's 159 members.

These realities are well-known to everyone involved in Space WARC. They have been obscured primarily because of the reluctance of the two sets of contending players to raise them publicly, presumably for fear of compromising their initial "hard" positions. A number of American studies (including one by an FCC industry advisory committee) have discussed the issue. By and large, however, the subject has tended to be given a secondary status in Space WARC planning exercises.

In reviewing the present role of INTELSAT and the other multilateral organizations in the Space WARC negotiations, this discussion paper suggests that the United States has a strong interest in actively examining a negotiating option that would give INTELSAT and the other networks an explicit role in any future ITU arrangements for frequency and GSO access.

Any expanded role for the multilateral organizations will require some scaling down of the present nation-oriented focus of both the northern and southern positions at the conference. Given strong sensitivities on sovereignty, this will be difficult to do. The alternative, however, could be a conference outcome which would impose regulatory conditions unfavorable to the steady current expansion of world satellite resources, to the detriment of all countries.

THE BACKGROUND FACTORS

To begin with, there is a specialized jargon in international telecommunications, as with any business. In order to complement other documents on Space WARC subjects, it is useful to adopt several specific phrases. In the ITU, member nations are referred to as "administrations." Multilateral organizations like INTELSAT are usually called "Common User Organizations." In this paper, to reduce prose clog, they will be referred to as CUOs.

In order to put the CUO issue in relationship to other Space WARC factors, it is useful to summarize two points: (a) the Space WARC process itself, and (b) the current procedures by which the ITU handles coordination of radio frequency and geostationary orbit (GSO) resources.

Space WARC conference activities have been divided

into three parts. The first is to consider the current situation for use of the geostationary orbit for communications satellites. The second is to decide what alternative arrangements may be necessary and for which frequency bands and services. Finally, the conference is to decide what principles and criteria should guide any alternative arrangement. This latter task will probably not be taken up until the second session of the conference in 1988. One certainty is that the Space WARC will modify a number of current ITU procedures. In order to understand the complexities involved in any changes the conference recommends, a brief review of the way in which access to GSO and frequency resources are handled under current procedures is in order.

The responsibility for this process has been assigned to an ITU component, the International Frequency Registration Board (IFRB). The board is a semi-autonomous unit within the ITU structure. Under the present system, ITU administrations submit requests for frequency and/or GSO resources to the IFRB for registration on its Master Register. The claim is honored if it conforms with established technical criteria, and if it is not challenged in terms of interference with a previously registered claim by another administration. The IFRB is not a regulatory agency in the normal sense of the term. Its role is to confirm or ratify the outcome of the registration process rather than to adjudicate or enforce any decision.

If a registration is challenged on the basis of harmful interference to a previously registered frequency or GSO slot, the matter becomes a subject of bilateral consultation between the concerned administrations. The IFRB may assist in the process but it is not designed to satisfy competing claims through enforceable regulatory sanctions. The system is porous enough that, in situations where an administration is clearly the offending party in an interference issue, it can insist that its claim be listed in the IFRB Master Register.

In terms of the Space WARC and its issues, it is important to note that the current IFRB system is not a structured planning process in the sense of identifying and enforcing optimal use of limited frequency and GSO resources. Its focus is on servicing one-at-a-time claims to a specific part of the resource. Resource conservation as such is not a factor. One result is that the IFRB Master Register contains many registrations that are unused or misused, complicating attempts to reduce congestion in international frequency use.

The primary beneficiaries for this so-called "first come, first served" system of registering frequencies and GSO "slots" have been the big satellite powers—notably the United States and the Soviet Union. The other big beneficiary has been INTELSAT, whose satellites carry the great bulk of international traffic. During the first two decades of satellite communications, there have been relatively few difficulties in obtaining available frequency and GSO resources. However, in two instances in recent years, two Third World countries—India and Indonesia—have had problems coordinating their satellites with those of INTELSAT and the Soviet Intersputnik network. These incidents, which were resolved, tended to reinforce LDC claims that it will be increasingly more difficult for them to have access to increasingly limited frequency and GSO resources as the big satellite powers continue to expand their present systems.

This will be the nub of the Space WARC debate during two sessions spread out over a three-year period.

At the present time, the CUOs are essentially outsiders to the debate. The ITU is an organization of sovereign states; the CUOs attend its conferences as observers. Their interests in ITU regulatory coordination are handled by individual states, known as notifying administrations. (The United States, and specifically the FCC, serves this role for INTELSAT.) There are also working contacts between INTELSAT and the ITU for coordination and other matters. Nevertheless, the essential point is that the CUOs currently have no direct administrative or legal representation within the ITU framework. Thus Space WARC interests will be decided by its members, acting individually or in regional or ideological groups. Although INTELSAT, in particular, has discussed Space WARC issues within its own governing bodies, there will not be an "INTELSAT caucus" at the conference.

Despite this arm's-length relationship with the ITU, the CUOs will be directly affected by any decisions the conference takes. Their organizational interests would probably be best served if the conference makes no significant changes in the present ITU procedures. The current system is flexible enough to give CUOs the GSO slots and frequencies they need with relatively few co-ordination difficulties. It is unlikely, however, that the present procedures will be left untouched. The more probable outcome involves some form of more structured planning and coordination process, with the possibility of preassignment of GSO slots and frequencies on a country-by-country basis. Whatever variation is selected, such an outcome would not be helpful to INTELSAT in particular. Rigid preassignment, from which it would be excluded by definition, could limit the present range of flexibility it enjoys in effectively planning and coordinating its GSO and frequency requirements.

Realistically, any preassignment plan will have to con-sider INTELSAT needs. This could involve, for instance, some form of arc-segmentation arrangement for its GSO requirements. What-ever accommodation was made, however, INTELSAT would be locked into a long-range planning system that could limit its ability to respond to options made possible by advances in satellite tech-nology or by its own changing operational needs. The result would be to limit capabilities for efficient aggregation of both its own services as well as GSO and frequency resources.

As suggested above, these prospects are directly rele-vant to preparation of U.S. proposals for the conference. INTEL-SAT and other CUOs represent an important factor in any viable middle ground between the current essentially open-ended system which benefits big satellite powers and the extremes of rigid a priori procedures which could tie up otherwise useful resources in an essentially political solution that would, at best, only partially respond to legitimate future LDC satellite needs.

How does the CUO factor fit into a workable U.S. strategy? Basically the United States seeks a viable formula that will continue to provide the flexible benefits of the present ITU procedures, adapted to LDC concerns about future access to GSO and frequency resources. It will, in particular, have to address alternatives for "guaranteed" access short of LDC resource-vesting proposals. There are a number of components involved here—

technical, political, and economic. This paper will discuss the CUO factors which are common to each of these in any overall U.S. strategy. The paper makes the following assumptions:

1. There are flaws in the present procedures in terms of providing sufficient assurances for practical access to GSO and frequency resources down the road for both new entrants as well as present operators. At a minimum, the United States will have to propose some adjustments in the present procedures to accommodate LDC concerns.

2. The LDC proposals for rigid a priori vesting of rights in these resources, country by country, lower the prospects for guaranteed access by reducing the overall ability to adjust the resources flexibly as overall access needs evolve.

3. The pragmatic guarantor of international access for most ITU members are the CUOs. About 120 of the union's members are also members of one or more CUOs. The remainder are, by and large, mini-states with little or no international traffic. In the case of INTELSAT in particular, equitable access is reinforced at three levels:

(a) *technical,* through planning procedures that consider the international and domestic needs of INTELSAT members in the design of advanced satellites and in their operational modes.

(b) *economic,* through efficient aggregation of GSO and frequency resources, together with tariffing procedures that favor smaller countries and profit-sharing arrangements that can help finance overall national telecommunications development.

(c) *political,* by providing each INTELSAT nation with an element of control over organizational decisions in the planning and operational process. Through weighted voting procedures, INTELSAT decisions are still dominated by a small group of industrial nations, the heavy users of the system. However, smaller nations have increasing influence, through aggregation of their shares by region (as provided for in the INTELSAT permanent agreements) or through direct pressure on the organization's plans.

4. In its Space WARC proposals, the United States will have to consider the role of the CUOs in any viable plan for future GSO and frequency coordination. Its options run from a contin-

uation of present procedures to proposals for giving CUO's a more direct role in the ITU coordination process. This paper examines the latter set of options.

5. Any proposals for giving CUOs a more direct role should be developed as realistic alternatives to current LDC a priori planning proposals. There is a range of options here. Their common theme will be to raise the "guarantee threshold" by involving the CUOs in the planning process in ways that take advantage of their ability to meet members' needs on an efficient collective basis. In effect, they would have some form of priority consideration in the coordination process, working out efficient patterns within and between CUOs. Their planning would carry special weight in the overall process because of their ability to aggregate GSO and frequency resources more efficiently.

6. Any such special consideration would not abrogate or modify the right of any ITU administration to register its own national requirements through the union's current procedures. The difference, of course, is that (under some formula) their needs would be considered in a "second round," after CUO requirements are submitted. The second-round procedure would involve coordinating overall CUO requirements with individual national requirements. The presumption is that most national requirements would be accommodated in the initial round.

Before looking at some of the CUO options available to American preparations for the Space WARC, it is useful to review background factors pertinent to any decisions in this area.

THE MISSING CUO FACTOR

The obvious question to ask is why hasn't the CUO factor received more attention? Collectively, they represent the largest single operational element in international satellite communications. INTELSAT alone handles more than half of all international traffic, if one includes cable traffic but excludes microwave and other regional traffic in North America and Europe. Despite this massive reality, INTELSAT and the other CUOs are effectively on the sidelines in any formal discussion of future international regulatory arrangements such as the one that will take place at Space WARC in 1985 and 1988.

The orthodox reason for this is the structure of the 125-year-old International Telecommunication Union. By custom and by treaty, it is tied to the fiction of the preeminence of national sovereignty in telecommunications matters. At a time when national boundaries have become increasingly less relevant to telecommunications, the sovereignty factor has been strengthened in the ITU. The reason for this, of course, is the value that the majority of the union's members, the smaller developing nations, put on their influence in a one-nation, one-vote organization. No plan that would cede significant powers to the CUOs, superseding the present distribution of sovereignty, would be acceptable. Any modification of ITU coordination procedures will have to accommodate to this fact.

Resolution 3 of the 1979 WARC, which recommended the Space WARC, focuses only on national access to resources. This is despite the fact that, in reality, no more than 10 percent of the union's member have, or can be expected to have, in the foreseeable future, need for direct access to GSO or frequency resources. In any event, there is no mention in the resolution of the role of the CUOs as a factor in any revised planning procedures.

This is not the result of mass amnesia about the CUO role, or any lack of understanding by the key players on the need to fit the CUOs into any planning process. The reasons are essentially political. The activist Third World countries which engineered the Space WARC resolution were primarily interested in keeping the focus on resolution language that would imply the need for some form of sovereign vesting of resources. Any mention of the CUOs would have deflected this focus. The United States and other big satellite powers attempted, with some success, to get language in the resolution that would not prejudge the planning method. Again, any special mention of the role of the CUOs would have deflected this focus. As a result, in the intense negotiating over the resolution language, there was no consideration of, or interest in, explicitly acknowledging the potential role of the CUOs in the Space WARC agenda. In summary, the CUOs and their interests will be a large and shadowy presence at both sessions of the Space WARC, the largest single satellite resource of a majority of the delegations but represented, as such, by none.

DEFINING THE COMMON-USER ORGANIZATIONS

Who are the CUOs? The answer would seem simple enough, but not in the current complexities of international satellite affairs. Defining the CUOs will, in fact, be a major element in any strategy for factoring them into a revised ITU planning mechanism.

The obvious definition of a CUO is an organization of two or more ITU administrations which jointly own and operate a satellite system for their international and/or domestic requirements. INTELSAT is such an organization. Several regional organizations also fit this definition, e.g., the Arabsat group.

There is a second definition. This is a satellite facility which is owned by, or under the regulatory control of, a single ITU administration but whose services are utilized by one or more other administrations under bilateral arrangements. The current example of this is the Indonesian Palapa II satellite. Palapa circuits are leased by Malaysia, the Philippines, and Singapore. Another variation on this are the several commercial proposals in this country to lease or sell satellite capacity for international operations in the North Atlantic region.

The distinction between these two types of CUO arrangements is important. This paper will restrict itself to a discussion of the first type, i.e., jointly owned and operated systems. It is quite possible that arrangements of the second type will become more common in the future. Given attitudes within the global telecommunications community, however, it is unrealistic to expect that the Space WARC conference would agree to any kind of special status for common-user facilities owned or regulated by a single administration. Any such proposal would be perceived as seeking preferential treatment by one set of administrations vis-à-vis the others. It would also be challenged by the orthodox CUOs.

Moreover, in terms of the specific focus on U.S. interests in this paper, there is a strong case against supporting preferential treatment, for several reasons:

1. Although the United States has an interest in encouraging commercial international satellite operations by American firms, advocacy of preferential treatment that might benefit these firms could be interpreted as a lessening of the U.S. commitment to INTELSAT. There is, moreover, no firm indication, now or in the future, that U.S. firms would need such protection.

2. The United States has little interest in encouraging the development of national satellite systems abroad which might adopt a strategy of leasing services (such as the Indonesians now do) to other countries in ways that may undercut INTELSAT.

3. The United States has an interest in encouraging smaller countries, particularly in the Third World, to continue to rely on INTELSAT for their international and domestic needs, or on jointly owned regional systems. In terms of their own self-interest, such jointly owned systems can provide developing countries with a wider range of services than national systems. More important, reliance on other national systems can involve a significant loss of control by a country over its own telecommunications. Participation in INTELSAT or a jointly owned regional system gives them some role in the planning and operation of the system.

This, in turn, forms a critical part of the strategic argument against the a priori planning proposals being advanced actively by those developing countries who are in a potential position to become regional satellite leaders, e.g., India and Brazil. It is relevant to ask the smaller developing countries whether their essential interests are best served by reliance on these regional "big powers" and, in particular, whether they are not better off with a modified ITU planning arrangement that gives more adequate attention to the needs of the INTELSAT system and other common-user organizations in which they have more direct control.

In summary, any U.S. proposals for giving greater recognition to common-user needs should be restricted to the inclusion of jointly owned multilateral organizations in any revised coordination planning arrangements.

THE U.S. INTEREST IN A CUO STRATEGY

In examining any strategy for an enhanced CUO role in ITU coordination procedures, it is useful to note that CUO and U.S. interests are not always the same. Historically, U.S. policy has been to support INTELSAT as its chosen instrument in international satellite affairs. Although the United States has only a 24 percent controlling share in the organization, it is the dominant voting power.

More recently there has been a small but significant shift in the long-standing policy of unquestioned support for INTELSAT's role as the monopoly global carrier. This shift has taken place with the proposed entry of commercial U.S. satellite carriers in intercontinental operations which will have some competitive effect on INTELSAT traffic. This debate, currently carried on in an intensified form, will have to be kept apart from any U.S. proposals for an enhanced role for regular common-user operations at the Space WARC. The best way to do this would be, as suggested earlier, to eliminate any nationally owned or regulated multilateral operation from consideration as an international common-user organization.

Within the U.S. government, several planning exercises have given specific attention to the role of the CUOs in space communications policy. They are a Congressional Office of Technology Assessment (OTA) study in 1982 analyzing the results of the 1979 WARC, and an FCC industry advisory group on preparations for the Space WARC. In addition, a May 1984 FCC Notice of Inquiry discusses the subject.

The OTA study suggested the need to plan world satellite resources on the assumption that domestic satellite capacity in most countries would probably be made available on a joint-use common-use basis through INTELSAT and regional arrangements. The study proposed greater policy attention to the role of CUOs in fashioning a viable overall satellite strategy.

The OTA report goes so far as to suggest that the United States and other developing countries should encourage privately funded joint ventures with developing countries to construct and operate regional CUO systems to meet their current domestic telecommunications needs. Such an approach, the report suggests, would offer the prospect of relieving the pressure on LDC support for an a priori planning regime: "If low cost and technically attractive domestic satellite capacity is made available through an international organization that accommodates the sovereignty interests of each country, many developing countries could come to see access to orbital slots and satellite frequencies as a side issue with availability of service being the main objective."

The relationship of the CUOs to the Space WARC is also discussed in a 1984 FCC industry advisory committee report on Space WARC planning. The committee reviewed possible U.S. approaches to integrating CUO needs with those of individual countries. Its comments are significant in reflecting an approach that is consistent with the overall U.S. goal of maintaining the flexible aspects of the current ITU resource-assignment mechanism.

The committee's Working Group C looked at a range of proposed planning methods which might be considered at the Space WARC. In evaluating middle-ground methods which could be acceptable to the United States and like-minded administrations, the group chose as first among the "preferred order" of planning methods a combination of "access demand planning" and "guaranteed access by means of multilateral coordination." Both of these methods are consonant with the concept of an enhanced CUO planning role discussed in this paper.

Working Group B of the committee conducted an intensive review of the legal and institutional factors involved in Space WARC issues. The institutional study, in particular, discusses the CUO role in any workable resolution of these issues.

The committee's January 1984 report drew upon these studies in making its major point that the "United States should be prepared to make concessions to preserve the essential advantages of the existing regime." It then goes on to discuss, in general terms, how this might apply to CUOs:

as a legal matter, the United States should be prepared to advocate the position that conflicts between individual states and common user institutions should be resolved consistently with any independent treaty obligations imposed by the charter of the common user organization. As a corollary of this notion, when the conflict exists between a common user system and a state or states that are not bound by its treaty, "equitable" access objectives might be satisfied by an accommodation that confers the greater good to the greater number of states. Alternatively, an arbitral procedure to arrive at an internationally refereed decision might be used. In these ways, common user systems might have rights regarded as equal to those of independent systems sponsored by individual Administrations acting outside of a common user framework. The United States could propose that such principles be integrated into the existing coordination procedures.

As noted above, current ITU procedures have been generally successful in permitting CUOs to provide a high degree of "equitable" and even "guaranteed" access to satellite services by their member-nations, i.e., the overwhelming users of international satellite communications. The FCC advisory committee report is correct in noting that ITU regulations involving the CUOs "provide a working reconciliation of the sovereignty notions that underpin the ITU with the collective decisionmaking that characterizes international organizations."

The FCC has also issued four Notices of Inquiry (NOI) in preparation for the first session of the Space WARC. (These notices are intended to invite public comment on policies and proposals currently before the commission.) The fourth and final Space WARC Notice, issued in May 1984, discussed, among other subjects, a possible planning role for the CUOs. Specifically, it reviews the option that ITU administrations could identify their future network needs through "different institutional settings." The Notice points out that these potential settings can vary in terms of their jurisdiction (world, regional, and sub-regional) and the kind of forum to be used. This could be, the Notice suggests, an ITU forum or a non-ITU multilateral body. This latter category could, of course, include INTELSAT and/or other common-user organizations, although these are not mentioned specifically. The Notice points out that some combination of one or more of these mechanisms is also possible.

The FCC document makes the important point that a

wide variety of multilateral facilities planning activities already exist. The United States participates in a number of these on a continuing basis in the North Atlantic, Pacific, and Carribean regions. There are comparable arrangements in other regions. Additionally, the two ITU technical consultative committees (CCIR/CCITT) have related planning exercises. Finally, INTELSAT engages in a similar identification process on a quarterly basis.

In the NOI comments on this subject, the commission says that it is not "unalterably opposed to the use of multilateral forums for the identification of satellite requirements." This is, obviously, a backhanded way of saying that it doesn't think much of the idea. Its preference (reflecting overall U.S. government policy to date) is to cite what it calls the "many compelling reasons for relying on the initiative of individual Administrations to unilaterally identify and describe their required satellite networks on a case-by-case basis as they arise, using the IFRB to disseminate the information."

The foremost reason for favoring this approach, the NOI states, is the complexity of the technical and operational aspects of designing and using satellites. Moreover, it notes, the subject is complicated by the range of domestic policies involved in each different country. The Commission's conclusion is that attempting to shift this procedure from its present focus on individual administration planning to a multilateral forum would lead inevitably to substantial difficulties.

There is no question about the soundness of the Commission's comments on this subject in terms of long-standing U.S. interests. This country has a highly structured system for processing governmental and private-sector GSO and frequency needs. The system is designed to operate effectively within present ITU procedures.

The essential point for any Space WARC strategy is that these procedures are going to be modified. Whatever benefits the LDCs—the ITU majority—now get from the present system of multilateral facilities planning activities, these activities are not perceived as enough to satisfy the "equitable" and "guaranteed access" standards set in the Space WARC agenda.

The extreme LDC position is to impose a strict planning

regime, involving predetermined "ownership" of GSO and frequency resources. To counter these views, the U.S. proposals must
be responsive to the essential elements of "equitable" and "guaranteed" access, while retaining a realistic measure of the present
flexible procedures. The current consultative arrangements described in the FCC Notice can be an important continuing part of
any such pattern. But, given the political situation at the Space
WARC, something else is needed. As this paper suggests, a closer
look at the role of INTELSAT and the other CUOs should be part
of any approach to a workable U.S. strategy.

It would, of course, be naive for the United States to
base its Space WARC strategy on the assumption that, if the present
system works, there is no problem. There is a problem as long as
the adoption of some sort of long-range a priori allocation system
is possible. LDC thinking on this subject was formed, in part, after
two leading LDC activists, India and Indonesia, had difficulties in
coordinating domestic satellite and GSO frequency needs with
INTELSAT and Intersputnik (the Soviet network) in the 1970s.
The fact that these coordination problems were resolved should
not obscure the equally important fact that both India and Indonesia had to make technical concessions which they regarded
as harmful to optimal efficiency of the systems they were planning.
These two examples will be cited repeatedly by Third World delegations as justification favoring a priori planning arrangements
at the Space WARC.

For the United States and other industrialized countries, the question is whether the present coordination procedures
as they affect the CUOs can be improved in ways that deflect such
criticism as well as provide a more viable basis for coordinating
both CUO and national needs. The thrust of this paper is the need
for a closer examination of a strategy which would give a greater
positive role to the CUOs in aggregating the resource needs of
their members as the initial step in the ITU planning and coordination process. Coordination difficulties between this CUO aggregation process and independent national requests could be resolved through an "arbitral procedure to arrive at an internationally
refereed decision" (in the words of the FCC advisory report). The
result would be (as the FCC report implies) a newly defined form

of equality between sovereign states and the CUOs in the procedures for sharing resources.

There are hazards for U.S. interests in this proposed process. As the largest single user of both domestic and international satellite facilities, these interests, potentially at least, are at risk in submitting to arbitration procedures that go beyond the current general formulations. There is a specific risk, directly touching on national security interests, if the process were to affect the considerable U.S. stake in military satellites. Any revised coordination formula would have to include assurances protecting the sovereign right of any country to obtain its basic satellite resource requirements. Given the strong proprietary interest that most ITU administrations have about these rights, any radical modification is not a likely prospect. Nevertheless, some modifications are implied in any formula that narrows the gap between the present unobstructed view of sovereign rights and the lack of CUO rights.

With these caveats, it is reasonable to assume that proposing some form of enhanced role for INTELSAT and other CUOs in an ITU satellite-resources planning process would be in line with basic American policy and interest. The hazards lie in the details of what may finally be decided at the Space WARC.

It is useful now to turn to an analysis of the present and potential attitudes of other countries toward proposals for more direct CUO participation in the ITU coordination process. These countries divide roughly into three groupings—the Third World, the Europeans, the Soviets, and the Chinese.

THIRD WORLD ATTITUDES AND ACTIONS

Space WARC is largely the result of an initiative by Third World countries to correct what they perceive is the imbalance in apportioning satellite resources. The initiative came largely from a small group of countries which had the technical expertise and the political will to force the issue at the 1979 general WARC conference. These countries were India, Brazil, Algeria, Indonesia, and, on the fringes, Yugoslavia. Most other developing countries played a very small role in the process, except to provide general support for the initiative.

There has been no significant questioning within the Third World of the need to revise present ITU procedures along the "equitable" and "guaranteed access" themes of the 1979 resolution mandating the Space WARC. Developing countries have often demonstrated their ability to vote their own interests in ITU conferences even when these interests conflict with overall Third World ideological appeals. On the key Space WARC issues, however, they can be expected to support (at least initially) a priori planning recommendations for meeting the "equitable" and "guaranteed" goals set out in the conference agenda.

Over and above the ideological appeals at Space WARC, developing countries will cite what they consider to be a major precedent in support of a priori planning. Specifically, they will argue that the United States and other satellite powers have agreed to similar procedures in past ITU conferences. Their major example will be the decisions of a 1977 ITU conference on direct broadcasting frequencies in which specific GSO and frequency resources were vested on a country-by-country basis. (The agreement did not initially cover the United States and other western hemisphere countries, which adopted a modified version of the 1977 agreement in 1983.) The analogy between the 1977 agreement and the a priori proposals that will be submitted to the Space WARC is, however, an imperfect one. The 1977 plan involved a single satellite service. It dealt with a common technical standard, as well as a technology that had not been actively put into service. None of these conditions apply to the complex series of satellite services that will be looked at in the Space WARC. Nevertheless, the 1977 precedent will be prominently cited as an example of the feasibility of a priori planning and vesting of resources.

In summary, the Third World majority will come to Space WARC with a strong bias in favor of replicating, on a larger scale, the a priori planning pattern adopted by the ITU eight years ago for direct broadcasting.

Given this background, the prospect for workable alternatives to a priori planning may seem dim. Any counterproposals will be treated with suspicion, including the CUO options discussed in this paper. The more vocal LDC leaders will contend that giving an enhanced role to the CUOs does not fulfill the conference mandate of guaranteed, equitable access. In particular, they will argue that it undercuts this mandate by giving preference to INTELSAT, an organization dominated (through weighted voting) by the United States and other industrial powers.

It is an appeal that will have a certain force. It can be answered by setting aside the monolithic implications of the term "Third World," and examining the varied interests and motivations of developing countries in the satellite field.

In satellite matters, the most visible group of countries were those which actively sponsored the 1979 Space WARC conference resolution: India, Indonesia, Algeria, and Brazil, among others. Their common interest is that they are either now regional satellite powers or have aspirations in that direction. Because of their early involvement in active satellite operations, they have a knowledgeable team of experts on the subject. They have been articulate, persuasive spokesmen for Third World initiatives within the ITU. However, these countries also have other, more parochial interests in their evolving role as regional satellite powers. Any proposal to strengthen INTELSAT (or potentially rival regional systems) within the ITU framework will probably be regarded by them as being against these interests.

The role of these countries at the Space WARC should not be minimized. They have a clearly defined sense of their own interest, and of its relationship to overall Third World concerns. Their message to other developing countries is an attractive one: establish your control over a critical set of natural resources in the one United Nations organization where developing countries collectively have a treaty-protected ability to do so.

The temptation for the smaller LDCs—the majority of ITU members—to accept this argument without question is strong. It involves the appeal of the free lunch, of getting something for nothing. It is an appeal that will be difficult to counter. The Western arguments emphasize technical objections to the a priori approach. However valid these arguments are, they do not add up to a successful strategy that will convince a significant number of LDCs to reexamine their generally unquestioned support for a priori planning.

A workable strategy will be directed to their broader

interests in satellite communications, well beyond technical details. Their interests lie in access to satellite services, not to GSO or frequency resources. Almost without exception, they depend on INTELSAT for their international satellite services. Increasingly, they also use INTELSAT facilities for a range of domestic satellite services. Over the next decade, more small countries will also depend on supplemental services supplied by regional CUOs. The prospects of developing their own individual satellite facilities are, in almost every instance, remote. Thus the concept of vesting rights in a package of GSO slots and frequencies, however attractive as an exercise in international pork-barreling, has little practical value.

The current Third World scenario, as put forward by a minority group of activist countries, is not responsive to these realities. Purely in terms of the economics of satellite systems, most LDCs will not be able to use their vested resources for discrete national purposes. The prospect of leasing these resources to other countries or to commercial ventures is a totally unproven alternative. The only possible Third World beneficiaries of an a priori assignment system would be a small group of larger countries (e.g., India and Brazil) whose populations and geographical mass justify a national system. As has already been demonstrated on a small scale in the case of the Indonesian Palapa satellite, smaller LDCs might benefit from concessional access to such national systems. The obvious disadvantage is that they would have no planning or management control, or hope of financial returns, in such an arrangement. In the not inconceivable circumstance of political crisis within their region, they could be cut off from access to a satellite wholly owned by a hostile neighbor, with predictable harm to their own national telecommunications facilities.

All this is by way of returning to the fact that their realistic prospects, now and in the future, lie principally with CUO arrangements as the best guarantee for equitable access to the services they need. This can involve INTELSAT and/or regional systems. In both instances, they have a management share and the hope of a profitable return on their investment.

In summary, the LDCs break down into two broad categories, measured by their realistic interests, as they prepare for the Space WARC. The small group includes countries which have now or will have in the future an interest in developing a national satellite system, with possible regional extensions. An a priori resource allocation process could, arguably, benefit them.

The large group—the majority of ITU administrations—includes countries who are out of the running in terms of developing national satellite systems. Their realistic interests lie in access to a range of services provided by CUOs. They are the countries that would benefit directly from an enhanced CUO role in the ITU planning and coordination process.

This suggests a convergence of interests on the future role of CUOs, moving toward a middle-ground resolution of the key Space WARC issue, one that could serve the interests of the LDC majority as well as those of the United States.

It involves, in broad terms, a planning and coordinating system that would establish a form of priority for the CUOs in identifying their GSO and frequency needs on a continuing "rollover" basis. Given the reality of INTELSAT's dominant role in global satellite traffic, the requirements of most administrations would be met first by coordination within INTELSAT, then by coordination with other CUOs, and finally, at the IFRB level, by coordination with those requirements of individual administrations which are not met in the initial coordination rounds.

There are clearly a number of loose ends to be tied up in any such arrangement. One of them involves the thirty or more mini-states which are not members of INTELSAT or a regional system. Many of them are, in fact, serviced by INTELSAT; special provisions could be made to have their interests represented by INTELSAT and/or a regional grouping.

Such a pattern would provide most LDCs with a co-ordination regime that would rely more, in terms of "guaranteed access," on their ownership and management participation in CUOs which are capable of the technical and economic aggregation of facilities that can give them, in reality, the full range of their required services.

For the United States and its industrial partners, it would mean ceding some precedence in the coordination process to CUO needs, without, however, surrendering their own individual right to access to GSO and frequency resources for national systems.

EUROPEAN AND CANADIAN ATTITUDES

The Europeans and Canadians share with the United States general opposition to the kind of a priori plans put forward by Third World activists. As a result, they are interested in acceptable alternatives. However, their receptivity to the idea of giving the CUOs a more prominent role in any planning process is less predictable.

Like the United States, the Europeans would be concerned that any such pattern not threaten their continuing plans for domestic and regional satellite development. The Canadians would be less concerned: they have an active domestic network, plus good working relations with the United States in regional satellite coordination.

The Canadians might be most receptive to a plan that gave an enhanced role to CUOs in the ITU. They have an instinct for this kind of compromise approach. The Europeans as a group might be somewhat more wary. In particular they will be mindful of the difficulties they had in coordinating their regional satellite arrangements with INTELSAT several years ago.

Second, the Europeans would probably weigh commercial considerations in any evaluation of such a strategy. The European satellite industry continues to plan a secondary role to the Americans, particularly in the key area of INTELSAT contracts. The Europeans will compete vigorously (helped by government subsidies) for the large number of satellites planned by INTELSAT, other CUOs, and by individual countries between now and the end of the century. As a result, the Europeans will consider the effect of these commercial prospects in any proposals for specialized CUO participation in overall satellite planning. With these caveats in mind, it is probable that the Europeans would be amenable to any strategy involving the CUOs that promises to modify the threat of a priori satellite-resource planning.

THE SOVIET AND CHINESE ATTITUDES

Soviet reactions to such a strategy are, predictably, more difficult to judge. The Soviets were more adamantly opposed to the calling of a Space WARC conference than any other industrialized country. They have, of course, an equal stake in heading off any a priori planning proposals.

Despite this, they followed their usual pattern of letting the United States and other industrialized states take the heat during the debate on the 1979 Space WARC resolution in Geneva. They would undoubtedly like to follow a similar course during the Space WARC, unless there was some indication of agreement early on in the conference on a viable alternative to a priori planning. Since this is unlikely, the Soviets will probably revert to their traditional posture of allowing the West to take the debating heat.

They are, however, realists in these matters. A strategy involving a great coordinating role for the CUOs would interest them. Their concerns would probably center around the status of their own common-user organization, Intersputnik. Traffic on their system aggregates to something less than one percent of INTEL-SAT's total traffic. For bargaining purposes, the Soviets might press for a formula that equates INTELSAT and Intersputnik—a fiction they attempt to sustain in various international forums. Nevertheless, their interest in any workable alternative to a priori planning is probably strong enough to override such a tactic. If an enhanced role for CUOs emerged as part of an acceptable alternative to a priori planning, the Soviets would probably support the proposal.

While professing ideological sympathy with Third World concerns over resource allocation, the Chinese have generally distanced themselves from specific endorsement of a priori solutions. They probably perceive their interests in this area as being closer to those of the Western countries. They have a major interest in expanding their domestic satellite network.

THE ATTITUDES OF INTELSAT AND THE ITU

Aware of the WARC's importance to the organization's future. INTELSAT has submitted several papers on the subject to its board of governors, providing details of the conference's relevance to INTELSAT operations. The board and the Assembly of Parties have not yet focused on the subject. One reason for this undoubtedly is that most INTELSAT members have not themselves defined their own detailed Space WARC plans. Specifically, they have not related their own national approaches to their INTELSAT interests. INTELSAT's strategy regarding Space WARC could be a significant element in the overall pattern of the conference, beginning with the first session next year.

Over and above the question of a possible enhanced role in ITU coordination procedures, INTELSAT will have several specific concerns. One will be its relationship to other international CUOs. INTELSAT is clearly the outsized member of this group and will continue to be so for the foreseeable future. How will its needs be weighed against those of Intersputnik or the smaller regional networks? Second, INTELSAT will be concerned about the status of nationally based CUOs, e.g., Palapa. Given the current dispute over a U.S. commercial entry into international satellite markets, INTELSAT can be expected to oppose any ITU recognition of such networks as legitimate common user organizations in a revised coordination plan.

The other organization with a stake in the Space WARC outcome is the ITU. The 120-year-old union is, among other things, an experienced bureaucracy, conditioned to resist change. This resistance is magnified by the fact that the organization does not have a unitary structure. Its component parts operate semi-autonomously, under a directorate-general which provides overall management guidance and support. The ITU element most concerned with Space WARC is the International Frequency Registration Board (IFRB), the agency that carries out the coordination procedures for all radio frequencies. The IFRB will be wary of any plan that appears to threaten its traditional prerogatives. An enhanced role for INTELSAT and the other CUOs in the coordination process for GSO and frequency registrations could be seen as such a threat.

However, the instinct for survival at the ITU is also alive and well. ITU officials know that a confrontational showdown at Space WARC, and the possibility of a failed conference, would be a serious threat to the union's future effectiveness. As a result, its officials have a stake in assisting the development of compromise solutions, including those that may appear to impinge on traditional ITU responsibilities.

IMPLEMENTING AN AMERICAN STRATEGY

Any U.S. strategy dealing with the Space WARC will be a combination of elements—political, economic, and technical.

There has been somewhat less attention to the overall political factors that will be as much of the conference environment as the technical.

The United States is not going to the conference to defend in toto the present ITU satellite-resource coordinating system. Such a defense would be self-defeating. There are good reasons for adjusting the system to new realities. If the United States and other countries with similar points of view cannot propose imaginative policies, other alternatives will be adopted by default. The result could be some form of rigid assignment plan and would be a step backward from the current workable, although imperfect, coordination process. Whatever its faults, the present system has been a critical factor in permitting satellite networks of all kinds to expand at a prodigious rate during the past twenty years.

U.S. policy is to secure agreements that maintain the essential flexible characteristics of the present system. The primary barriers to achieving this end are not technical or economic. They are political.

This paper has outlined the reasons for giving more attention to the CUO factor in the U.S. proposals. Such an approach offers an opportunity to moderate a large share of LDC concerns about future access to GSO and frequency resources. In developing this subject within a U.S. strategic framework, the purpose should be to establish, as a procedural matter, the level at which all ITU administration needs can be efficiently and equitably aggregated by giving some precedence to CUOs in a new form of ITU planning and coordination process.

The process would not preempt the right of any administration to register its own national requirements at any point in the coordination cycle. The primary constraints would be those already in force. They provide that coordination problems between individual administrations and CUOs should be resolved consistently with any independent treaty obligations imposed by the CUO charter. The FCC industry advisory committee has suggested a useful corollary covering disputes between a CUO and an administration not bound by the CUO treaty. In such cases, the committee's report proposes that equitable access objectives might be satisfied by an accommodation that confers the greater good to the greater number of states. Alternatively, the committee report notes, an arbitral procedure should be established.

"In these ways," the committee report concludes, "common user systems might have rights regarded as equal to those of independent systems sponsored by individual administrations acting outside of common user frameworks. The United States could propose that such principles be integrated into the existing coordination procedures."

These procedures would have to be worked out carefully. By way of example, one option would be a three-step procedure for bringing the CUOs into the planning and coordination process.

The first step would be an institutional arrangement within the ITU which would specifically acknowledge the role of INTELSAT and other CUOs in a planning cycle for the coordination of future frequency and GSO requirements. INTELSAT and other CUOs would have priority in preparing their requirements, based on projections of their current operational patterns. This planning coordination would take place under ITU auspices between eligible CUOs. The procedure could also include ITU administrations who are not members of a common user organization, but who might elect to have their needs included in the CUO planning exercise. In addition, administrations which operate, or plan to operate, national systems could participate so that any requirements independent of their CUO involvement could be considered in the overall aggregation of needs.

This planning cycle would be on a "rollover" basis. As a result, there would be relatively limited adjustments at any one point in the process over the years. The purpose of the exercise would be to accommodate, to the widest extent possible, the domestic and international satellite services needs of all administrations through common-user systems. This would provide a pragmatic substitute for equitable treatment and guaranteed access in an a priori system. The critical difference would be the enhanced ability of the CUOs to (a) aggregate technical and economic resources in ways that service their members more effectively and (b) to conserve frequency and GSO resources.

The second step would be to submit these jointly coordinated CUO plans, with related registration requests, to the IFRB under the present notifying administration procedures or, possibly, directly. These submissions would form the base for the IFRB's overall satellite frequency and GSO registration process. Provision would have to be made for the contingency that the CUO submissions do not cover (a) all individual administration requirements and (b) CUO requirements which, when submitted to the IFRB, did not resolve all technical compatabilities between CUOs or between a CUO and an individual administration.

This resolution process would take place in a third step through (in the words of the FCC industry advisory report) "an arbitral procedure to arrive at an internationally refereed decision." The details of these procedures will have to be carefully studied. On the one hand, it should include safeguards against any arbitrary restrictions on national satellite development. On the other hand, the procedures will have to be strong enough to satisfy LDC administrations that, potentially, their right to practical guaranteed access to frequency and GSO resources (via the CUOs) will be protected against so-called "first come, first served" preemption by national systems. Developing an acceptable consensus between these two requirements will be a difficult but critical part of the acceptability of any workable arbitration procedure.

RECOMMENDATION

There is a strong case for more active consideration of the role of CUOs in a revised ITU coordination procedure for space frequency

and GSO resources. An enhanced CUO presence in these procedures could be an important step toward narrowing the current
gap between the perceptions of North and South groupings at the
conference. If this approach is a viable one for the United States,
the next step is to develop a set of specific proposals for inclusion
in the overall U.S. Space WARC proposals. Under ITU rules, these
are scheduled to be submitted by February 1985. At the same
time, it will be important to consult with other administrations
to test the proposals, both in terms of content and the degree of
support they can be expected to receive in the conference itself.

S I X T E E N

Latecomer Cost Handicap: Importance in a Changing Regulatory Landscape

HARVEY J. LEVIN

At the Second Unispace Conference in 1982, a first Draft Report submitted by several developing countries proposed that certain developed countries vacate the lowest space satellite spectrum—the C-band—and move up and develop the higher KU and still higher KA bands. Those developing countries perceived this as a means to shunt the higher development, capital, and operating costs of systems installed in higher frequency bands, onto the advanced nations best able to bear them. This would supposedly leave the more congenial lower frequencies, which are less expensive to operate in, to the less affluent nations.[1] The proposal was subsequently withdrawn through a Brazilian initiative in the face of strong resistance by developed countries (DCs) determined to protect their investment equities.

More generally, Third World resentment of the practice of awarding rights to build space satellite systems on a first-come, first-served basis, is seemingly based on what those nations perceive as the dwindling availability of slots or orbit spectrum assignments. The developing countries (LDCs) also fear the handicaps they suffer due to the higher R&D and engineering costs incurred to open up new bands at higher frequencies.

Very similar issues have in fact been raised within the United States. In television, e.g., UHF stations are more costly to build and operate than VHFs, in that, at a signal quality comparable to that enjoyed by VHF stations, they require far more signal power per 1,000 TV homes reached. Furthermore, latecomer VHF entrants must also pay far more to buy an existing VHF as congestion grows than they would have had to spend to build a new one at the outset.[2] Similarly, AM radio latecomers had to buy existing AM stations in the market at inflated prices when that band became too saturated to accommodate new entrants. Furthermore, AM latecomers also had to pay a lot to avoid illegal interference with incumbent AM licensees in congested regions—hence the high coordination costs which latecomers incur, to which we return momentarily.

There is a third and final domestic analogy in land mobile radio where, once again, capital and operating costs appear to rise notably per square mile of area covered, as we move up from 50 to 800 MHz.[3] This seems to be true whether we hold power constant and increase antenna heights, or hypothetically hold antenna height constant and increase power. Here, too, in this important radio band, latecomer firms appear to suffer clear cost handicaps. But again, the handicap is not only technical.[4]

Within this framework my paper examines five issues: First, it offers an overview of the equity and efficiency problems in the management of global spectrum resources. Second, it identifies latecomer cost handicaps and then analyzes their consequences in theoretical terms. Third, it outlines a methodology through which to test the hypothesis empirically. However, empirical studies could not be undertaken here, nor even summarized. Fourth, the paper considers the degree, if any, to which free rider benefits enjoyed by latecomers may at least mitigate their alleged cost handicap. Fifth, it briefly identifies no less than seven arrangements through which net latecomer cost handicaps (assuming such do exist) could be avoided or offset, perhaps significantly, and concludes by describing an eighth, somewhat different approach.

The serious review of such arrangements should arguably figure in United States preparations for Space World Administrative Radio Conference 85–88 (ORB85). Such a review could provide the basis for important U.S. initiatives in that far-reaching conference. However, there is little evidence that such initiatives, or even the necessary preliminary research and analysis, are now being undertaken in any systematic fashion.

AN OVERVIEW: EQUITY VERSUS EFFICIENCY
IN THE MANAGEMENT
OF GLOBAL SPECTRUM RESOURCES

"Parking places" in the geostationary orbit, and the space satellite frequencies associated with them for purposes of information delivery, constitute scarce communications resources. Guaranteed equitable and efficient access to the same constitutes a kingpin of policy imperatives in the International Telecommunication Union and other international organs today. Orbit spectrum is by no means unlimited in its current or projected availability, but its precise degree of scarcity is still subject to debate by engineers, technologists, lawyers, and economists. However, identification of the services, spectral bands, and orbital regions, access to which is already congested, or seems likely to become so, is both possible and an urgent item of business for the first session of Space WARC85.[5]

Among a large number of conceivable organizational, procedural, and policy approaches for managing these precious resources, the dialogue in international decisional arenas has until recently focused increasingly on two polar bounds. First is the notion that the evolutionary process now in force will best serve the interest of all nations. It would allegedly do so by facilitating an economically efficient use of orbit spectrum consistent with flexible responses to changing telecommunications needs in a context of technological change, all on a first-come, first-served basis.[6] Second is the notion that some form of detailed a priori planning will best serve the interests of all nations in the equitable use of global orbit spectrum resources.[7]

Critics of our present coordination procedures further observe that, under the principles of first-come, first-served, those procedures impose sizable (and for the critics, unjustifiable) cost burdens on developing countries as latecomers or new entrants into space satellite communication.[8] Stated otherwise, early arrivals are said to shunt off extra costs on latecomers even though orbit spectrum is deemed by many to be a global resource belonging to all nations, and not just to the few nations with the technology and know-how to use it now.

For the critics of detailed a priori plans,[9] on the other hand, those planning mechanisms will unavoidably operate to stifle or freeze the kind of technological advance and flexibility deemed to be highly essential to implementing the ITU directive that orbit spectrum be managed with economic as well as technical efficiency.

Finally, in developed countries as well as LDCs, there is evidence that a far wider range of planning options are under review, and that the initial rigid dichotomy between first-come, first-served and detailed long-term plans has begun to break down.[10]

THEORETICAL ANALYSIS OF LATECOMER HANDICAP

For analytical purposes I assume that firstcomers (in advanced economies) may enter prematurely for fear that rivals would do likewise. I further assume this may result in a land-rush syndrome such that there will be: (a) uneconomic excessive entry by developed country satellite entities, but also more competition in those markets on that count at least; (b) latecomer cost handicap, and hence impeded entry for developing countries, domestically and internationally, when the latter are otherwise ready to enter such fields as high frequency radio, land mobile, terrestrial TV broadcasting, or fixed and broadcast satellites; (c) a priori, pre-engineered planning of direct broadcast satellites by LDCs to reserve orbit-spectrum assignments for themselves, thereby further intensifying the DCs' rush to enter, and even to innovate for narrower spacing of satellites in the orbital arc even if uneconomic.

In this latter regard, the question is: Do latecomers also impose costs on incumbents, by forcing them to narrow their spacing by using more sophisticated, expensive equipment toward that end? Or does the prevailing ITU legal-administrative-regulatory doctrine tacitly operate to impose all coordination costs on the latecomer who must in fact accommodate to the firstcomer incumbents?

My main focal point here is on the second and third issues. However, I propose also to assess the contention that technical advances by DCs, and resultant technology transfer, will act to reduce satellite capital and the operating costs at least enough to mitigate latecomer private cost handicap. A supplemental issue necessarily excluded, due to time and space limitations, relates to the optimal size management unit for orbit spectrum resources generally, with special reference once again to the level of entry barriers which face new entrants.

In particular, then, I want to focus on telecommunications market structures, all on the assumption that orbit spectrum location has discernible effects upon entry barriers with special reference to the cost handicaps of latecomer users (firms, governments, and other entities), as well as to the tendency toward premature entry, investment, and occupancy by users in advanced economies.

A number of surprisingly neglected issues for review here are: (a) the allegation that the DCs claim-staking strategy will result in fuller, faster entry than otherwise, even if uneconomic; (b) the further allegation that latecomers may be relatively if not absolutely precluded from orbit spectrum, due in large part to significant private cost handicaps, albeit partly offset by compensating free-rider benefits where the DCs innovate new technology; (c) the final allegation that LDCs press for detailed a priori pre-engineered allocation plans, so as to prevent serious entry-blockading cost handicaps, and to safeguard Third World access prospects, even though such plans may blockade DC entry in certain regions and services, and also force DCs to innovate uneconomically for narrower orbital spacing.

In regard to the relative position of latecomer and firstcomer users of orbit spectrum and other information resources where users may include firms, governments, and governmental agencies, the question is whether information systems suffer cost handicaps in using higher spectral regions, elliptical or random orbits, or in designing narrower orbit spectrum spacing. And whether latecomers therefore suffer such handicaps across countries as well as within their own. Or from still another viewpoint, does the uncertainty of any future access at all (or at least the fear of cost disadvantages due to less preferred locations) lead latecomers to seek detailed a priori planning and earmarked access rights, all at the expense of economic efficiency and flexibility for latecomers and incumbents alike?

In terms of a theoretical a priori microeconomic model, the divergence of private and social cost under current spectrum management arrangements must be discussed relative to three separate factors. First, the apparent zero price of spectrum to users permitted access to it under the present centralized nonmarket system of allocation on a first-come, first-served basis. Second, the extra cost imposed on potential (next-best) users where, as latecomers, any prior usage denies them access to that spectrum in whole or in part, or reduces signal quality due to congestion and higher interference levels. Third, the lower resultant private than social cost of spectrum usage which will then be related to latecomer cost disadvantage by considering the process whereby entry continues until spectrum congestion and interference raise marginal private spectrum costs to a level where the demand for and supply of spectrum come into equality.

Microeconomic analysis then further enables us to hypothesize about latecomer cost disadvantage as it may in principle operate within nations at various stages of economic development, or across nations competing for limited orbit and spectrum resources.

TOWARD AN EMPIRICAL ASSESSMENT OF LATECOMER COST HANDICAP

To assess the latecomer hypothesis empirically, I am developing a data base to probe the links between spectral location on one hand, and cost levels on the other, in particular as they affect entry

barriers and market structures in international space satellite communications and the U.S. domestic land mobile radio service. In
addition, attention will be paid to the differences in component
costs, power requirements, energy expenses, and other operating
costs in television broadcasting, for licensees using low VHF channels, high VHF, or UHF channels. I will also examine the resultant
impact on several sets of transfer prices of station facilities traded
in TV markets, all as the price of latecomer entry.

In each case, I want to test the hypothesis that, ceteris
paribus, the higher the spectral region (or radio frequency) within
any service, the higher the capital and operating costs for any
spectrum users. But a second objective must also be to determine
whether there is quantitative evidence that latecomers benefit from
cost-reducing innovations by firstcomers as much or more as they
(the latecomers) suffer from handicaps in having to operate higher
up. At issue there are so-called free rider benefits and "learning
by use," to which I turn later. But a further word first on eco-
nomic-engineering evidence.

In moving up from C to KU-band and again to KA,
there are increasing problems caused by rain attenuation. As a
consequence, more and more signal power is needed to sustain
the same signal quality and area coverage as in a lower band. In
going from C to KU, e.g., there could in principle be a 6 to 1 cost
increase due to the 6 to 1 power increase necessitated. Even mitigated by so-called modulation improvement, the power and hence
cost increases could still be as much as 3 to 1.[11]

True, if we accept a lower degree of reliability—say,
99.7 percent instead of 99.9 percent—we could reputedly reduce
cost increases in the move from C to KU to a doubling only. But
how much reliability will the LDCs at most be willing to give up?
The issue is a delicate psychological-political one in part because
the LDCs already resent what they view as second-class service
in KU, subject as that is to loss of coverage due to heavy rainfall
in the tropics where many of the LDCs are located. Just as they
also resent being relegated to lower cost but preemptible INTEL-
SAT transponders in the domestic leases they hold; and just as
they have long sought access to satellites to escape sunspot-in-
duced distortions in their use of HF broadcast spectrum. These

LDC demands are no less intense even though preemptible, but lower-cost INTELSAT circuits may well be "all they really need" today, and even though the aberrations of HF radio may also be cost-effective though seemingly unreliable and of poor quality.

Some of the cost handicap that latecomers incur may actually reflect the relatively greater scale economies, which equipment manufacturers for C-band now enjoy; the relatively smaller scale economies at KU, where the demand for equipment is still small, though this may of course change in the future; and the much larger nonrecurring R&D costs incurred in developing new bands like KA. Once again, this leaves the LDCs sensitive to (and resentful of) their status as latecomers.

To test this latecomer cost hypothesis in space satellites, a number of preliminary statistical models are being developed and will soon be applied to a set of satellite cost data. These models are sketched briefly below, along with a few for land mobile radio. But let me turn first to the origins and incidence of so-called coordination costs, themselves a function of orbital and spectrum congestion, and current administrative-legal-regulatory practice.

Latecomer Coordination Costs

By coordination costs, reference is normally made to the locational and power costs incurred by a latecomer to avoid interfering with an incumbent user. Strictly speaking, it could also refer to the incumbent's extra cost in accommodating latecomers with minimal extra costs imposed upon them.

Even at lower C-band frequencies, then, latecomers are disadvantaged when these bands fill up and become congested. Hence the origin of latecomer cost handicap lies in (a) the harsher propagation characteristics of higher spectral bands, and (b) the far smaller scale economies in producing new equipment for the newer, less fully utilized higher spectral bands. But these cost handicaps may be offset in part by the lower coordination costs in the less congested higher bands than in the more congested lower bands.

At some point, then, this will undoubtedly lead to latecomers choosing between higher coordination costs in the lower congested bands (though with lower equipment costs due to better propagation conditions and larger scale economies), and lower coordination expenses but higher nonrecurring engineering costs in the higher, newer bands. What my several estimates of latecomer handicap could well reveal, then, may be (a) costs that rise as increased power is needed to offset poor propagation conditions, while holding constant (b) scale economies in equipment manufacturing, and (c) nonrecurring engineering plus R&D costs. Last, my latecomer cost estimates also hold constant (d) the incidence of coordination costs as between latecomers and incumbents, and across the several spectral bands, reflective of their varying degrees of congestion.

The Case of Space Satellites
        To test this latecomer cost hypothesis in space satellites, a number of statistical models could be examined. One model might first regress real cost of the space segment in 1984 dollars, on the number of "equivalent transponders" in each satellite, the assumption here being that, ceteris paribus, the greater the satellite capacity, the higher is the absolute cost (though not necessarily cost per transponder). Second, we could regress real cost on design life of satellite in years, the assumption being that, ceteris paribus, longer lived equipment would be more costly in absolute terms. Third, we could regress real cost on the frequency band in which these transponders operate. Last, to capture the full impact of band differences on cost, we might well interact the band dummy variables with the number of transponders.
        A good place to start, then, is to fit equations with the following variables:

Y = spacecraft costs in 1983 dollars,
X-1 = dummy variable equal to 1 if KU band, zero otherwise,
X-2 = dummy variable equal to 1 if KA band, zero otherwise,
X-3 = number of equivalent transponders,
X-4 = satellite design lifetime in years,
X-5 = number of equivalent transponders times dummy variable equal to 1 if KU, zero otherwise,

X-6 = number of equivalent transponders times dummy variable equal to 1 if KA, zero otherwise,

X-7 = number of years design lifetime times dummy variable equal to 1 if KU, zero otherwise,

X-8 = number of years design lifetime times dummy variable equal to 1 if KA, zero otherwise.

To calculate the impact of adding one more equivalent transponder in KU rather than C, or KA rather than C, I would next calculate the interacted coefficients for KU and KA. These would enable us to estimate:

(1) the impact of one more equivalent transponder on cost, if in KU and if in KA,

(2) the impact of a switch from C to KU, and from C to KA, at the mean level of transponders,

(3) the impact of one more year of design life, if in KU and if in KA,

(4) the impact of a switch from C to KU, and from C to KA, at the mean level of design life.

An alternate model might then cast our dependent variable as real cost of spacecraft per equivalent transponder as such, or per year design life, and then regress simply on band differences and the number of equivalent transponders. The coefficients would then reveal the impact of a switch from C to KU, and C to KA, on real cost of spacecraft per indicated divisor. That is, the most revealing dependent variable would be real cost per transponder year (adjusted for design life), or better still, per circuit year, assuming 1,000 voice circuits per transponder.[12]

Needless to say, many statistical problems must be resolved in developing either model—problems of functional form, multicollinearity, simultaneity bias, etc. Nonetheless, the variables specified in each model here do serve to illustrate a place to initiate the kind of empirical assessment that could help us determine the validity of Third World contentions that as latecomers they do suffer significant cost handicaps, albeit offset in part perhaps by innovational benefits they enjoy from firstcomer investment, occupancy, and use of orbit spectrum resources.

Time must in any case be introduced as an independent variable in some suitable form, perhaps by using pooled models for time series data. For INTELSAT, in any case, investment costs per circuit year declined from $32,500 for INTELSAT I (in 1965) to $662 for INTELSAT VI (in 1986), whereas real cost per transponder year declined from $12,480,000 to $440,000.[13] By the same token, real costs per transponder year for U.S. domestic satellites declined, 1972–1982, from $500,000 to $280,000.[14] Crude industry estimates are that perhaps half of this cost decline is due to innovational advance (extending transponder capacity and design life), and half due perhaps to greater familiarity with the technology in use (learning curve).[15]

Nonetheless, there is virtually no systematic published analysis of the impact of band location on real satellite capital and operating costs per transponder year, or per circuit year. Indeed, the most one can discern from the above sources is that, among the twenty-nine U.S. Fixed Service domestic satellites reported in Lovell and Fordyce,[16] real costs per transponder year for four KU-band SBS satellites (av. = $930,000), have unit costs over twice the unit costs of 25 C-band satellites (av. = $386,000, with only Spacenet operating a hybrid C/KU band satellite).

In contrast, Pelton reports the depreciated capital cost per transponder year of fixed domestic C-band satellites to be $300,000–$350,000, compared to $350,000–$400,000 for an international Fixed Satellite service "with expanded coverage and interconnectivity." The latter is deemed more likely to use KU-band equipment,[17] which means, at worst, a cost penalty of 30 percent in the switch from C to KU.

Beyond this, Lovell and Fordyce estimate that a direct broadcast satellite, necessarily operating in KU, will have "a figure of merit of better than $2 million per transponder year."[18] Yet DBS and Fixed Service satellite costs cannot be compared accurately because channels in each case vary considerably in number. Nonetheless, the switch of a Fixed Service satellite from C to KU band raised power requirements (a major cost component) 3 1/2-fold, and another 5-fold for a KU-band Broadcast Service satellite.[19]

The above discussion relates to the space segment alone,

including launch costs and insurance. However, there is reason to believe that ground segment costs may also be higher, ceteris paribus, for latecomers forced to enter the KU-band, or eventually KA, in the face of growing C-band congestion. This, too, needs close scrutiny.[20]

The Case of Land Mobile Radio

Are the trade-offs similar in land mobile radio? To reach the same size area using equipment in different land mobile bands, in principle the range of cost increments are proportionate to the needed increments in antenna height or transmission power. For argument's sake, let us hold antenna heights constant, say, at 200 feet, and assume flat terrain, a base/mobile radius of 30 miles, and zero man-made noise in the area. Then, in rising from the lowest land mobile band at 50 MHz, to the highest at 800 MHz, rough hypothetical engineering estimates are for a needed power increase from 35–75 watts at 50 MHz, to 150–200 watts at 450 MHz and 800 MHz. This is roughly a six-fold increase over the whole range of land mobile frequency bands.

By the same token, holding power constant at 75 watts, and again, assuming flat terrain, no man-made obstructions, a base/mobile radius of 30 miles, and zero man-made noise, to cover the same square mileage antenna heights would ideally have to rise from 110 feet at 50 MHz, to 250 feet at 150 MHz, 475 feet at 450 MHz, and 650 feet at 800 MHz. Once again, this is a six-fold increase in antenna size over the full range of land mobile frequencies.

Transmission power and antenna heights are two major cost components in land mobile radio systems. Assuming, unrealistically to be sure, so-called mid-line mobile systems, operating on one frequency and one channel, the very crude hypothetical cost estimates on the assumptions just stated are as follows:

System Band	System Cost
50 MHz	$800
150 MHz	850
450 MHz	1100
800 MHz	1400

The less than doubling of cost notwithstanding, our hypothetical increase of antenna heights and power by six-fold may in part reflect the fact that cost-effective, real-world systems normally vary both radiated power and antenna heights, not just one such input.

The question is why latecomers do not complain, even when they must enter higher bands with higher capital and operating costs. Why, indeed, when new equipment costs will also be higher because scale economies are smaller at 450 and 800 MHz than at the older lower 50 and 150 MHz bands. The answer may be that in higher bands, besides higher costs, there is also greater quality (no "skip"), more reliability, and wider coverage as benefits. And also that, over time, the cost of next-best non-spectrum alternatives (fuel, vehicles, drivers) rises substantially, and so, too, does the magnitude of cost savings in land mobile use, even at higher bands.

But do latecomers need such benefits? If not, the higher equipment costs would be hard to swallow without resentment and the latecomer may in fact be forced to pay more than the firstcomer, albeit for more benefits, but unwanted benefits at best.

There are in fact additional factors which make for more costly antennas in the higher spectral regions. Reference is made here to the cost of scarce antenna sites, often on high buildings in congested urban environments, as, e.g., the Sears Tower in Chicago or the World Trade Center in New York City. There is, then, a real estate issue in that locational rents are appropriated by landlords and these raise the cost of the antenna site even more than otherwise. Hence, in crowded urban centers, latecomers will move up from 50–150–450 MHz, and again to 800 MHz. They do so to insure signal reliability and high signal quality in the face of man-made noise and artificial obstructions. Under those conditions, high antenna towers are increasingly important and, with them, much higher antenna costs, including equipment and special sites.

Without probing all necessary details, an illustrative equation from which to initiate an estimation of the impact of mobile radio spectral location on the area covered by given transmission power and/or antenna heights, would be the following:

Y = square mile area covered in 75 mile radius
        from population center of SMA,
X-1 = transmitter power,
X-2 = height above average terrain of antenna
        (HAAT), including where possible height of
        antenna tower,
X-3 = dummy variable for terrain equal to 1 if moun-
        tainous, zero otherwise,
X-4 = dummy variable for terrain equal to 1 if near
        water, zero otherwise,
X-5 = dummy variable for terrain equal to 1 if con-
        gested with large urban structures, zero oth-
        erwise,
X-6 = frequency level used by base station,
X-7 = transmitter power × frequency level (X-1 ×
        X-6),
X-8 = HAAT × frequency level (X-2 × X-6).

Once more, my hypothesis is that (a) lower bands fill up first, (b) higher bands, though less congested and less subject to technical vagaries of their own than the low bands, require equipment with higher power and/or antenna heights, than in lower bands. Therefore, the higher band equipment should presumably be more costly, other things being equal, a tendency which would be further underscored insofar as (c) scale economies are larger in the lower, older, more congested bands than in the less congested higher, newer bands for which equipment production is still limited.

The biggest problem in any such estimation of latecomer cost handicap in mobile radio is that (a) any really large sample requires the use of the FCC's massive but notably imperfect data base; (b) detailed equipment cost data are not in any case available there, or readily available elsewhere except in crude broad-based estimates. My present model therefore (a) works without explicit cost data; (b) specifies only the two major system cost components for a land mobile base station, viz., transmitter power (in kilowatts) and HAAT (height of antenna above average terrain); (c) interacts each of these variables with frequency level to permit an estimate of the full impact of spectral location. A major additional factor further modifying the coverage area of any base station in reality is (d) character of the terrain, whether mountainous, flat, near water, or marked by urbanized structures, to capture which it is necessary to devise dummy variables that distinguish between major classes of terrain, using the FCC's digitized terrain map.

The most reliable and complete data records are for each base station's transmitter power (ERP). Records for antenna height above average terrain (HAAT) are far less so, except for the 470–490 MHz band where that information is recorded explicitly (including height of the antenna site too). Elsewhere it would be necessary to use antenna heights from base of site to tip, plus elevation of site above sea level. (The existence of high towers on skyscraper sites can also be detected by devising a dummy variable to distinguish between differing degrees of clustering of base stations within SMAs. Where high building rooftops provide the antenna site, one would in fact expect to find a significant clustering of transmitters.)

A few suggestive pieces of empirical evidence drawn from well-known radio engineering data support my approach. Together this evidence indicates that land mobile signals will be delivered such that any given service area requirement can be met by varying combinations of antenna heights and transmitter power. The higher the one, the lower the other may be, with visible constraints imposed by frequency band location, and notable differences between suburban and urban environments and their respective man-made noise and artificial obstructions.

Thus in table 16.1, at the upper end of land mobile frequencies (851–866 MHz), for any given antenna height, a 20-mile service area radius will require some 2–3 to 4–5 more effective radiated power (ERP) in urban environments than suburban. This implies that the more congested the area with man-made noise and obstructions, then the more costly the operation in terms of required antenna height and transmitter power (cf. var. X5).[21]

This is still better indicated in table 16.2—Radio Range Estimator Chart—where, e.g., at the upper extreme, antenna heights

TABLE 16.1. Equivalent Power and Antenna Heights for Base Stations in the 851–866 MHz Band Which Have a Requirement for a 32 km (20 mi) Service Area Radius

Antenna height (ATT) (feet) (meters)	Effective radiated power (watts)[a]	
	Urban/trunked	Suburban
Above 5,000	65	15
4,501 to 5,000	65	15
4,001 to 4,500	70	29
3,501 to 4,000	75	25
3,001 to 3,500	100	30
2,501 to 3,000	140	35
2,001 to 2,500	200	50
1,501 to 2,000	350	80
1,001 to 1,500	600	160
501 to 1,000	1,000[b]	220
Up to 500	1,000	500[c]

SOURCE: FCC Rules & Regulations: Part 90 (abridged), para. 90.635, table 2, p. 59.

[a]Power is given in terms of effective radiated power (ERP). Applicants in the Los Angeles, Calif. area who demonstrate a need to serve both the downtown and fringe areas will be permitted to utilize an ERP of 1 kw at the following mountaintop sites: Santiago Park, Sierra Peak, Mount Lukens, and Mount Wilson.

[b]Stations with antennas below 1,000 ft (AAT) will be restricted to a maximum power of 1 kw (ERP).

[c]Stations with antenna below 500 ft (AAT) will be restricted to a maximum power of 500 W (ERP).

rise to 15,000. There, at the left side, ERP is determined by varying levels of antenna height and transmitter power, such that at 10 watts, 25 watts, 50 watts, and 100 watts, comparable ERP generated by the above input combinations could ostensibly deliver information a full 55 miles using the 30 MHz band, but only 50 miles at 50 MHz, 30 miles at 150 MHz, about 21 miles at 450 MHz, and just a scant 19 miles at 800 MHz (cf. X6).

Thus, whatever ERP is generated by the several combinations of transmitter power (10w–100w, var. X1), and antenna heights (15'–15,000', var. X2), on the table's left-hand side, the distance that information can be delivered is smaller, the higher the frequency band used (cf. X6). Or stated otherwise, insofar as firstcomers ostensibly have access to the older, lower spectrum regions first, they enjoy the cost savings associated with the smaller

TABLE 16.2. Radio Range Estimator Chart

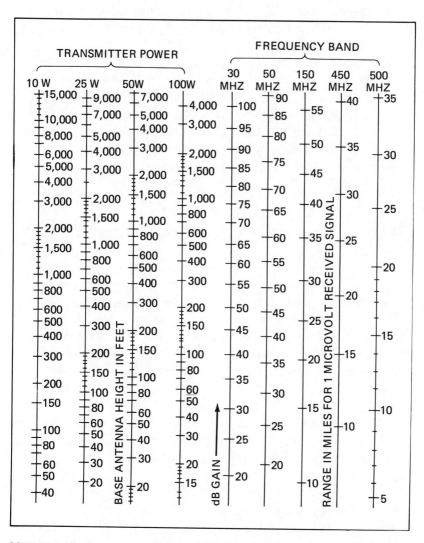

SOURCE: Midland Land Mobile Radio Co., Inc., "Answers to Most Questions Asked About FM Two-Way Radio," p. 13 (derived generally from *Reference Data for Radio Engineers,* 4th ed., ITT; see especially ch. 24).

antenna heights and transmitter power needed there to cover any given service area radius (than at higher frequency levels).

By the same token, latecomers constrained to move up to less congested higher frequency band locations must pay for higher antenna heights and/or transmitter power to cover the same service area radius.

In sum, the latecomer cost handicap we hypothesized about in space satellites and land mobile radio does seem consistent here with the basic propagation curves which underlie table 16.2, our Radio Range Estimator Chart. Together, finally, tables 16.1 and 16.2, and the other cited tabulations from FCC Rules & Regulations, pt. 90, underscore the case for testing the latecomer hypothesis with the far larger data base on actual land mobile radio records.

FREE RIDER BENEFITS AND THE LEARNING CURVE

A brief word next on free rider benefits and the learning curve which are said to mitigate latecomer costs. That argument is that latecomers enjoy free rider benefits from the costly and risky nonrecurring R&D which DC firstcomers perform in opening up spectral bands which LDC latecomers subsequently enter, too, without themselves having to design and develop the newer equipment. Per unit equipment costs are also alleged to be lower on that score than they would be without the firstcomer's growing familiarity with the technology in question.[22]

On the other hand, there is contravening evidence that once the firstcomer enters, he appears to secure a decisive advantage over latecomers who may follow, sustaining his gains over time. We don't really know why this should be so, but it is so. According to this argument, therefore, it is by no means clear that the firstcomer's growing experience with his technology will transmit any net advantage to the latecomer, or significantly impair the firstcomer's initial advantage.[23]

Spence's cogent distinction between firm-specific and industry-wide (interdependent) learning curves is clearly germane here. Assuming INTELSAT to be the incumbent, and LDCs the latecomers, firm-specific learning effects can indeed create substantial entry barriers to latecomers, with moderate rates of learning creating the greatest barriers.[24] In that case, latecomers will have greatest trouble in entering, their relative costs being larger than where incumbent's learning is zero or negligible. By the same token, it is at best uncertain that incumbent will suffer from excessive entry and resultant low output (experience and learning), and that its incentives to invest in learning or innovation must therefore be seriously impaired. Accordingly, latecomer cost handicap would not seem likely to be mitigated by even sizable firm-specific learning effects.

In contrast, industry-wide or interdependent learning curves may, under certain conditions, cause "more entry to occur . . . than is ideal," and, with such spillover effects, "competition (will be) increased and market performance improved, despite the reduced incentives for investment in learning . . . for . . . firm(s)." Hence "the industry-wide cost-reduction process (for given output rates)" will become more efficient, even though such interdependence, by reducing "the return to investing in (learning as) accumulated volume," will act to slow down the pace of cost reduction. (*Ibid.*, pp. 66, 67–68.)

For the analysis here, then, there may indeed be industry-wide learning effects which could mitigate latecomer cost disadvantages, the question being mainly one of (a) relative size, or scale economies, in each case; and (b) the frequency with which such postulated conditions do in fact recur.

Accordingly, at least two sets of hypotheses need to be tested; first, that moving to higher bands in the radio spectrum acts to raise capital and operating costs due to physical propagation problems, deficient scale economies in equipment manufacturing, etc.; and second, that while the firstcomer's learning curve and innovational advance may help offset some latecomer handicaps, they by no means need impair the firstcomer's net competitive advantage.

A final related issue relates to the contention that narrower orbital spacing offsets what LDCs would have enjoyed had they entered earlier, lower down. Today, e.g., the latecomer cannot get older cheaper antennas designed to work with wider orbital spacing, unless they are literally hand-tailored. So what LDCs resent here is having missed the earlier windfalls to the first-comer—before orbital spacing was narrowed, but after the scale economies in equipment production had materialized. Yet the firstcomers did risk venture capital, and spent a lot on non-recurring R&D.

Latecomer entry may be expensive, then, but if technical innovation helps bring down per unit costs of satellite technology for all countries, the LDCs may also enjoy some compensatory measure of free rider benefits. If so, the LDCs may be gratuitous beneficiaries of the communications activity of advanced country enterprises. The question is whether these compensatory benefits do or do not offset the entire cost handicap of latecomer entrants, no easy matter to determine. Much depends on whether free rider benefits to LDCs would significantly blunt the DCs' incentives for optimal research, development, and use of orbit spectrum resources.[25]

Just as traditional common pool problems may notably result in DCs doing too much production, investment, entry, and exploitation too soon, so the problem of LDC or DC free riders may result in too little entry, too late. One classic example relates to public goods generally, with benefit externalities so diffuse and hard to trace that not all who enjoy them can be charged directly, or refused access if they do not pay. Another example relates to DC technological advances costly and risky to bring about, but which could be readily and cheaply copied by imitators but for patent protection without which innovators are reluctant to proceed.

Using a priori deductive microeconomic models and secondary sources in the technology transfer literature, finally, one could examine such issues as these:

(1) Insofar as free rider benefits that lower entry costs may also hamper innovation, the question is which effect will prevail, and indeed, whether these two effects are even measurable, let alone separable.

(2) Can they in any case be reconciled—enough lag and appropriation for multinational corporations (MNCs) to retain incentives to innovate new technology, but quick enough transfer

(diffusion) of old technology for LDCs to enjoy lower entry barriers than otherwise (via cost reduction)?[26]

(3) Can the MNC literature on technology transfer further help us assess the case of, say, an LDC hiring a DC concessionnaire to develop the LDC's orbit spectrum? When and how, if at all, could DCs control the training, know-how, and software they must (or inadvertently may) leave behind with an LDC, and hence avoid losing their exclusive access to the technology? Could a DC avoid unwanted technology transfer here, i.e., compared, say, to when working through global or regional consortia of users?[27]

(4) Under what kind of arrangements, in what kind of industries, do license, consortium or joint venture agreements, or various kinds of contractual terms emerge? And where might these impair or facilitate technology transfer to LDCs, with or without disincentive effects on DC innovation?

POTENTIAL POLICY OPTIONS
TO OFFSET LATECOMER COST HANDICAP

A final word now on eight possible options to mitigate the net latecomer handicap left even after partial mitigation by free rider benefits.

The first seven such options entail the use of economic incentives, or mechanisms, that permit latecomers to improve their relative position in orbit spectrum usage, not only by acquiring and activating unused assignments before they are all occupied (nos. 6 and 7), but also by acquiring rights to access existing satellite systems, operating on someone else's assignments. Actually, such systems would be accessed either by using their idle capacity (options nos. 2, 3 and 5), or by drawing on their fully utilized capacity (nos. 1, 4, and 5). Only the final option (no. 8) is geared to recapture orbit spectrum rents generally and to defuse them widely among all nations, rich and poor alike.

First, a latecomer could presumably buy out another country's or firm's lower cost C-band satellite system. But the problem here is that an LDC may not really need a whole system, and that it is at best hard to find one that will precisely fill its requirements anyway. Still more important from an equity standpoint is that such a purchase would inevitably enable incumbents to collect orbit spectrum rents as firstcomers, all at the latecomer's expense.

So it may be better, second, to buy into a consortium, like INTELSAT or Eutelsat, operating in part at least, in a lower satellite band. Participation there would also permit the LDC to enjoy scale economies through the sheer magnitude of the service provided, at the same time benefiting from equipment designed to operate in the lower spectral region.

Or third, latecomers could just lease extra unused lower band satellite capacity, and hence avoid having to build their own system higher up. Or avoid having to develop narrower orbital spacing to squeeze themselves in.[28]

Fourth, latecomers could aim to alter the present coordination practice so that there would be a new, more equitable cost-sharing between latecomer and incumbent. This would presumably modify the de facto forcing of latecomers to bear the whole coordination cost themselves, under current arrangements, when they now try to enter a crowded band.[29]

Fifth, the latecomer could buy transponders outright from a private or common carrier-owned satellite facility capitalizing on lower spectral band location.[30]

A sixth option would be that of a joint venture where a poor LDC with limited resources provides an orbit spectrum assignment, and a DC entrepreneur invests the needed venture capital, with both sharing the revenues, much as when Rupert Murdoch agreed to finance a broadcast satellite system to be built on an assignment held by Luxembourg, in a joint venture between Great Britain and Luxembourg.

Seventh, and a variation of the above, is when an LDC could hire a DC firm to build, operate, and manage a system on the former's orbit spectrum assignment, much as where the Arabsat consortium hired Comsat for a similar purpose.

Eighth and last, we could work through ITU to enable all nations to recover orbit spectrum value as Ricardian rents which firstcomers can capture at C-band, rather than at KU or KA.

Among the mechanisms to diffuse these benefits widely among rich and poor nations, I am scrutinizing several to auction off use rights, recovering the related economic rents and earmarking them to develop telecommunications infrastructure in the Third World. One approach just short of this is simply to devise empirical techniques to estimate the value of orbit spectrum assignments as a factor in international negotiations between DCs and LDCs in an intergovernmental framework.

In this latter regard, the analysis of latecomer cost handicap might indeed help us develop such an empirical technique. When faced, say, with a 3 to 1 cost increment in operating space satellites at the higher KU rather than the lower C-band, and then another such increment going up to KA, the hypothetical latecomer should rationally be willing to pay something just short of this cost differential for access to the lower band. In principle, that is, latecomers should pay a sum just short of the cost savings C-band would have facilitated relative to KU, or KU relative to KA.

The spectral rents would be the sums that incumbents could hypothetically collect by selling assignments or systems to latecomers. Or, from another viewpoint, a lump sum tax would be the optimal device to recapture these rents, ideally without distorting resource allocation, or creating disincentive effects on R&D and technological advance. At present, however, there is unfortunately no institutional mechanism for levying or collecting such a tax.

A sometimes proposed alternative to such a once-and-for-all impost would be for a periodic but automatic diversion of value into some form of telecommunications infrastructure development. Automaticity raises a number of puzzling problems. These are not insuperable, but they do require careful analysis on the merits. Although widely opposed in the United States, automaticity does now figure prominently in the Highway Trust Fund and, a few years ago, in well thought out proposals by television's Carnegie Commission II. In these cases, user charges and royalty fees collected from highway users and commercial TV licensees are or would be earmarked for highway development and public broadcast service, respectively.

Nor, outside the United States, should we forget proposals for automatic annual diversion of some percentage of world GNP, or of global military expenditures, into Third World infrastructure development.[31] Nor, finally, analogous mechanisms for similar purposes within the International Monetary Fund, viz., that of Official Development Assistance, Special Drawing rights, and a Common Commodity Fund, all geared ostensibly to ease special development-related balance of payments difficulties of Third World countries.[32]

Note also that a political constituency must be built for automaticity. This could be done by promising that recaptured orbit spectrum rents would be used to buy telecommunications hardware, say, from big U.S. companies or other transnational corporations, and would be so used consistent with First Amendment values. To safeguard the latter might entail the bolstering of multi-voice media structures in Third World countries, plus the funding of a specific cultural-educational facility, say, for hemisphere-wide programs or news services. So much, then, for steps widely discussed in the spectrum field to make concrete the Common Heritage of Mankind Doctrine there at least.

CONCLUSION

Latecomer cost handicap arises in large part from such factors as these. First, the propagation characteristics of the more recently developed, higher radio frequencies in such services as space satellites and land mobile radio. Second, the costly increases in transmission power and facilitating electronic equipment required to overcome the impairment of signal quality and information delivery to which latecomers are normally forced to turn in the wake of firstcomer saturation of the lower, less expensive spectrum bands in specific services. Third, the higher equipment costs associated with newer spectral bands where the economies of large-scale manufacturing have not yet been fully realized. Fourth, the high non-recurring R&D and engineering costs incurred to open up newer spectral regions. Fifth, the higher coordination costs incurred by latecomers who must bear the cost of adjusting their coverage patterns and equipment design under current legal-administrative practice.

These cost handicaps are presumably offset, though only in part, by the benefits that latecomers derive from the firstcomer's cost-reducing innovations and learning curve. Furthermore, here, too, once the latter enters the field, he appears to gain an irretrievable advantage over latecomers which follow.

Accordingly, U.S. space conference preparatory efforts should direct far more attention than hitherto to conceivable arrangements to mitigate possible net cost handicaps of latecomer entrants, a disruptive issue that could well surface again at future space conferences.

NOTES

1. See Draft Report of the Second United Nations Conference on the Exploration and Peaceful Uses of Outer Space, U.N. Doc. A/Conf. 101/10/PC/L. 20/Add. 1, 2 (1982). Latecomer developing countries urged that "the newer and more expensive technologies that lead to better utilization . . . should be adopted by the developed countries and by international systems, so that the comparatively simple and cheap technologies (e.g., lower frequency bands 4 and 6 GHz) are freed for use by developing countries (para. 275). Significant cost reductions for LDCs would also be facilitated if "developed countries shift their satellite . . . systems to a different frequency band (e.g., 11/14 GHz), leaving the 4/6 GHz band basically for use by developing countries (para. 150).

2. See Levin, *The Invisible Resource* (Baltimore: Johns Hopkins University Press, 1971), pp. 219–228; also *Fact and Fancy in Television Regulation* (New York: Russell Sage Foundation, 1980), pp. 120–123, especially tables 4.5, 4.6. It is said that the VHF band could have been expanded, but that firstcomer incumbents wanted entry costs to be high for new entrants.

3. See generally Levin, *The Invisible Resource*, pp. 205–214.

4. For economists, latecomer cost handicap in outer space appears to be little different from what occurs with the settlement of a new agricultural area. There, where land tracts vary in fertility, early arrivals will claim the most fertile land, leaving less fertile tracts for the late arrivals. Under those conditions, the price of land will be set by the higher relative costs of cultivating the poorer tracts, and the early arrivals will eventually recover economic rents, or unearned increments (windfalls). Such rents are basically generated by the spread between the market price of land, and the firstcomer's lower relative costs in cultivating it.

5. See Harold G. Kimball, "Implications for the Future of Satellite Communication," presented at 1984 Annual Conference of International Institute of Communications, Berlin, Sept. 21–23, 1984, pp. 2–3 (hereafter called 1984 IIC Berlin Conf.).

6. A general review of first come, first served under current coordination procedures of the International Telecommunications Union (ITU) appears in the First Report of the Advisory Committee for the ITU's World Administrative Radio Committee on the Use of the Geostationary Satellite Orbit and the Planning of the Space Frequencies Utilizing It, Dec. 1983, sec. 4D (The Current ITU Arrangements Ensuring Access to the Geostationary Orbit), pp. 4-18 to 4-28 (hereafter called First Advisory Report).

7. This position was recently expounded by T. V. Srirangan, Wireless Advisory to the Government of India, Ministry of Communications, New Delhi. See his "Equity in Orbit: Planned Use of a Unique Resource," 1984 Berlin Conference for the Planning of the High Frequency Broadcast Band—A Viewpoint," 1983 IIC Annual Conference, Aruba, Sept. 24–27, 1983, pp. 2–5. For a further contrast of current first come, first served arrangements and planning alternatives at the imminent 1985 Space Conference, see varying perspectives of William H. Montgomery, "Views of the 1985 Space Conference—Perspectives on Frequency Planning in the ITU," 1984 IIC Berlin Conf., pp. 1–6; and Kimball, ibid., p. 106.

8. See Srirangan, "Equity in Orbit: Planned Use of a Unique Resource," 1984 IIC Berlin Conf., pp. 3, 4, 7.

9. This is a term of art, referring to a preplanned or preengineered assignment table that determines beforehand which users of orbit spectrum may operate where, using which frequencies and slots, for what purposes.

10. For a highly illuminating "devil's advocate" case for a priori planning generally as the way to mitigate latecomer cost handicap, see FCC Space WARC Advisory Committee, Interim Report of Proposal and evaluation Test Group, Nov. 20, 1984, SWAC/DOC PEG #8, secs. 1–3A; see also secs. 3B–3C. More specifically, the widest range of management arrangements appear in the eleven methods identified and reviewed in First Advisory Report, secs. D, pp. 5-19 to 5-35. These were indeed reduced to seven at the CCIR's Conference Preparatory Meeting in Geneva, June 25–July 20, 1984, including five methods laid out in the Report of IWP/1 contained in CPM Conf. doc. no. 30, a sixth submitted by the USSR (on loosely packed plans with minimum and maximum values for each assignment), and a seventh by China (focused on computerized decision making). See Report of the CCIR Conference Preparatory meeting (CPM), International Telecommunications Union, Geneva, June 25–July 20, 1984, Part 2, pp. 132–156. Most recently, the International Regulations Working Group of the Space WARC Advisory Committee fashioned an illuminating "Combined Planning Approach," drawing upon "concepts as-

sociated with all of the Planned Methods identified in the CPM Report." These include computerized modelling methods, the French interference hormonization method, the establishing of certain minimum-maximum values of technical parameters for flexibility, a basic 30-year a priori plan to implement, in which there would be an 8–10 year Consultation Conference, and a set of interim arrangements including simplified use of current coordination procedures. (See Second Advisory Committee Report for ITU WARC ORB 85, sec. 2 D4).

11. One interesting question here is whether spectrum allocation should in fact take climate into account, allocating the higher frequencies to arid çountries where rainfall is scant (e.g., Libya or Algeria). Would this conceivably facilitate lower cost equipment design even at higher frequency levels?

12. See Robert R. Lovell and Samuel W. Fordyce, "A Figure of Merit for Competing Communications Satellite Designs," *Space Communications and Broadcasting— An International Journal* (April 1983) (1):57.

13. *Ibid.*, pp. 57, 60. This may superficially appear to constitute an offsetting latecomer advantage. But cost differentials across different spectral bands continue to operate at every point in time, in the face, i.e., of declining investment costs per transponder year. Thus latecomer cost handicap seems most appropriately measured in cross-sectional analysis, at a point in time with technological advance and learning by use held constant, statistically.

14. *Ibid.*, p. 59 and figure 9; see also Joseph N. Pelton, "Satellite Telenets: A Techno-Economic Assessment of Major Trends for the Future," *Proceedings of the IEEE, Special Issue on Satellite Communications Networks,* November 1984, figure 3, p. 1,447.

15. See, e.g., comment on "Learning Curves and Yields," in *International Competitiveness in Electronics,* Office of Technology Assessment of United States Congress, November 1983, pp. 76–77; also, Pelton, "Satellite Telenets," especially pp. 1,450–51.

16. Lovell and Fordyce, "A Figure of Merit for Competing Communications Satellite Design," table 2.

17. Pelton, "Satellite Telenets," p. 1,449.

18. Lovell and Fordyce, "A Figure of Merit," p. 63.

19. *Ibid.*, table 3, p. 62.

20. See, generally, David H. Staelin et al., *Satellite Network Architecture: Technology Issues,* July 20, 1982 (MIT Research Electronics Laboratory), fig. 17, p. 52; table 1, p. 54; and figure 22, p. 83. Staelin also laid out an engineering-economic analysis of total systems configurations (space satellite plus ground segment plus terrestrial linkages). These are actually configured for all three bands—C, KU, and KA—though more for engineers than economists. Staelin, especially ch. 3.

21. This is further confirmed in FCC Rules and Regulations, pt. 90 (abridged), tables 3–4. Once again, those tables provide estimates of antenna heights and radiated power for suburban and urban environments, separately, but this time for service area requirements which decline from 0-mile radius to 5-mile radius or less. (Cf vars. X1, X2, X7, X8.)

22. This would be due to the rise in cumulative average output per man week with the cumulative growth in aggregate output, or, better still, in "cumulative gross investment (the total output of capital goods) as an index of experience." Kenneth J.

Arrow, "The Economic Implications of Learning by Doing," *Review of Economic Studies* (1962), pp. 156–157. During World War II, e.g., a doubling of cumulative airframe output acted to reduce direct labor requirements by 20 percent. Armen Alchian, "Reliability of Progress Curves in Airframe Production," *Econometrica* (October 1963). There do in any case appear to be cost advantages in being first to enter a new industry. Robert H. Smiley and S. Abraham Ravid, "The Importance of Being First: Learning Price and Stability," *Quarterly Journal of Economics*, (May 1983), 98(2):353–357.

    23. *Ibid.*, pp. 360–361.

    24. See A. Michael Spence, "The Learning Curve and Competition," *Bell Journal of Economics* (Spring 1981), 12(1):57–62. Admittedly, of course, international telecommunications is not really a competitive regime. But with the advent of new private entrants (such as Orion), and regional consortia and domestic systems, all potentially interconnected by future "intersatellite links," new competitive elements may eventually be infused into international telecommunications.

    25. From still another viewpoint, in my agricultural land analogy (note 4 above), the free rider problem could be conceived as technological change which reduces the cost of cultivation. There, if new technology acts to lower costs equally for all classes of land tracts, fertile and less fertile, the late arrivals will enjoy no offsetting benefit, and the firstcomer's relative advantage remains constant. In contrast, if technological advance were to reduce costs on the poorer lands relatively more, than the late arrival would enjoy some offsetting advantage. That is, the new technology would have made his tracts more comparable, in quality and productivity, to the more fertile tracts.

    26. This key issue is implicit in the discussion in R. Hal Mason, "The Multinational Firm and the Cost of Technology to Developing Countries," *California Management Review* (Summer 1973), 15(4):5–13.

    27. See generally, R. Hal Mason, "A Comment of Prof. Kojima's 'Japanese Type versus American Type of Technology Transfer,'" *Hitotsubashi Journal of Economics* (February 1979).

    28. See generally, Levin, "The Political Economy of Orbit Spectrum Leading," in 1984 Michigan Yearbook of International Legal Studies, *Regulation of Transnational Communications* (New York: Boardman, 1984), pp. 41–70.

    29. See veiled allusion to such a conception in William H. Montgomery, "Views of the 1985 Space Conference," pp. 3–4. (Robert Tritt presented the paper in Berlin, and further expounded the concept in the ensuing panel dialogue.)

    30. The issues here are concisely stated in Levin, "Regional versus Global Strategies in Orbit Spectrum Management—The Global Commons Revisited," presented at the December 1982 meetings of the American Economic Association, New York Hilton, in n. 27 and associated text. In its most recent form, latecomers could lease or buy transponders outright, and also enjoy free transponders to distribute public service programming. (See Application of Pan American Satellite Corporation for a Subregional Western Hemisphere Satellite System, before the Federal Communications Commission, May 31, 1984, part I.) Specifically, Panamsat proposes to sell or lease its transponders on a private, non-common carrier basis to international news organizations, television networks, and to countries and other entities, thereby providing video and radio programming, video text, electronic mail, telephone service, data and computer communications, etc. It would

also provide free transponder capacity to a 21-nation Organization of Iberoamerican Television to encourage free exchange of its programming throughout the Western hemisphere.

31. See *North-South: A Programme for Survival,* report of the Independent Commission on International Development Issues under the chairmanship of Willy Brandt (Cambridge, Mass.: MIT Press, 1980). See especially, pp. 242–245.

32. *Ibid.,* pp. 209–220. For a more detailed analysis see John Williamson, "SDRs: The Link," in Jagdish N. Bhatwati, ed., *The New International Economic Order: The North-South Debate* (Cambridge Mass.: MIT Press, 1977), ch. 3.

PART V: SOVIET SATELLITE COMMUNICATIONS

SEVENTEEN

Cooperation and Competition in Satellite Communication: The Soviet Union

JOHN D. H. DOWNING

According to *Krasnaya Zvezda*, January 24, 1980,[1] the first decision to pursue research into a space flight with cosmonauts was taken in the USSR in 1946. This is perfectly plausible, given Stalin's well-known predilection for prestige aviation projects.[2] We do not yet know much of the intervening stages, but in April 1965 Premier Kosygin formally proposed to other heads of socialist states that they jointly collaborate on scientific research into a variety of space topics. The proposal issued eventually in a major meeting in Wrocław in 1970 at which the title "Interkosmos" was formally adopted. The project took international legal shape on March 25, 1977, and now involves as members Bulgaria, Cuba, Czechoslovakia, East Germany, Hungary, Mongolia, Poland, Romania, the USSR, and Vietnam.[3] Thus seven European countries are involved, and three Third World countries.

THE INTERKOSMOS PROGRAM

Interkosmos has undertaken a variety of scientific research projects of major proportions: space exploration, remote sensing, biology and medicine, and meteorology.[4] Satellites have also been launched within the program by nations other than the USSR, such as the Czech Magion satellite and the Bulgaria 1300 satellite.[5] Cuba can now access the scientific data bases of the CIIST (International Center of Scientific and Technical Information of the Comecon Countries) via an Intersputnik satellite, and was involved in the collection of oceanographic data through Interkosmos-21, along with East Germany and Mongolia.[6]

The most spectacular aspect of the Interkosmos program has been its space flights with binational crews. To date there have been binational flights involving a German (DDR), a Czech, a Pole, a Bulgarian, a Hungarian, a Romanian, a Cuban, a Mongolian, a Vietnamese, and an Indian.[7] Additionally, the Cuban, Lt.-Col. Arnaldo Tamayo Mendez, was black.

(The second, as well as the first woman in space, was a Soviet citizen, as was the first woman to accomplish a walk—not the version of history embraced by all U.S. media, it might be noted.) An all-woman Soviet space crew has been forecast by Soviet sources for the fairly near future.[8] The hundredth person in space, Viktor Savinykh, was a Soviet citizen. As of 1980, the USSR had had twice the number of human space-flight-hours as the United States.[9] These last-named aspects of the space program are more Soviet than collaborative, but they add up to an important total picture of venture in space exploration, and in communication about, as well as via, space. Out of the specific Interkosmos program had emerged by late 1981 no less than 500 public joint scientific papers.[10]

SOVIET-FRENCH COOPERATION

Joint scientific space projects between France and the Soviet Union began from 1966.[11] Nineteen eighty-four marked their twenty-first bilateral conference, which meets on alternate occasions in each country. The most visible form of their cooperation was their joint space flight in 1982, but this was only one aspect of their collaboration. In 1979, France spent 27.62 million francs on space cooperation with the USSR, and in 1980, 32.33 million; by contrast, it spent in those years 24.6 and 38.355 millions on collaboration with the United States.[12] Despite this relative shift between 1979 and 1980, the figures indicate comparability.

The research projects involved have been various.[13] They have included the VENERA experiments (exploration of the atmosphere and surface of Venus), INTERBOL and Arcad 3 (magnetosphere experiments, especially the costly Arctic Auroral Density study), ASTRON (astrophysics in the ultra-violet domain), SIGMA (experiments with X-rays and gamma-rays in space), and COMET (meteor-dust collection). In the rest of the present decade, other projected joint experiments are a satellite for the lunar polar orbit, a solar study, further work on Venus, and the AELITA study of the distance and movement speed of galaxies. In 1985 a satellite will be launched—with U.S. involvement as well—containing white rats and rhesus monkeys. And in 1986 there will be two Franco-Soviet satellites observing Halley's comet—in competition/collaboration with the European Space Agency, Japan, and NASA. The Franco-Soviet study will utilize as a key element a spectrometer developed jointly by France, the USSR, Hungary, and Bulgaria. On the French side these experiments are coordinated by the Centre Nationale des Etudes Spatiales, on the Soviet side by the Academy of Sciences.

SOVIET-INDIAN COLLABORATION

India is the other most favored non-bloc nation for Soviet space collaboration. Out of the six satellites actually sent up by India (i.e., not including the use of NASA for the SITE educational TV experiment in 1975–76), four have been made in India and launched from the USSR (the other two were European and U.S. cooperative projects). The Soviet-Indian satellites contained a mixture of scientific instruments from both countries. The work involved has included studies of the upper atmosphere, research in gamma astronomy, monitoring of satellites, and remote sensing. As noted, in April 1984 there was a joint Soviet-Indian space flight. One of the experiments involved a bio-medical study of yoga practiced in weightless conditions. The collaborative remote sensing satellite, IRS-1, was planned to survey both the land and the oceanic periphery of the subcontinent.[14]

Particularly noteworthy was the sequence of events in 1982–83, when a Ford Aerospace communications satellite for India's use failed comprehensively, losing at least $20 million for India. The USSR gave India free access to one of its own Intersputnik satellites for nearly a year, through which to broadcast, until a new satellite could be successfully launched. It is a matter of some dispute how far the numerous facets of the ground segment were actually in place to utilize this gift's full potential, but of its symbolic value there can be no doubt.[15]

MARITIME SATELLITE COLLABORATION

Joint activity in this field has taken two forms: involvement with Inmarsat, and participation in the COSPAS-SARSAT project. The Soviet Union is the second most active member of Inmarsat, the 40-nation maritime navigation satellite agency set up in 1979 with its headquarters in London. After the United States, whose stake is 23 percent, comes the USSR with 14 percent. The first Soviet ground station for Inmarsat was completed in Odessa in 1983. In July the same year it was reported that the USSR was offering for sale to Inmarsat a modified Proton rocket for the third (Marecs) satellite launch, at a quite competitive price—approximately $24 millions, as opposed to $28 and $30 millions respectively for its U.S. and French competitors. As a further carrot, it was offering a second launch if the first should fail, at half price.[16]

COSPAS-SARSAT is a satellite system designed for maritime and other distress situations and rescue operations (though occasionally Inmarsat has been deployed for these as well).[17] The nations and agencies involved are Morflot (the Soviet merchant marine ministry), the Centre Nationale des Etudes Spatiales, the Canadian Department of Communications, and NASA. Norway has also been involved in some aspects of the project. Salvage operations have been enabled as well. By the end of July 1984, two hundred lives had been saved.

THE INTERSPUTNIK SYSTEM

Insofar as attention has been given to mass communication by satellite, the organization most discussed by far has been Intelsat, broadcasting to nearly 170 countries, and with 108 members.[18] Yet the Intersputnik system, with current links to Algeria, Bulgaria, Cuba, Czechoslovakia, East Germany, Hungary, Iraq, Poland, Afghanistan, Laos, Mongolia, Vietnam, and the USSR itself, and impending links with Syria, Nicaragua, South Yemen, and Kampuchea, is certainly worthy of attention.[19] Point-to-point communications are an important feature of the Intersputnik system, but are presently not discussed in western or eastern sources that I have been able to discover. Here, therefore, I shall focus on television, with occasional references to radio.

The origins of the Intersputnik system lie in the Khrushchev era. In January 1959 the Technical Council of the USSR Ministry of Communications placed in the Seven Year Plan some basic proposals for a cosmic retransmitter of TV programs in color. In June 1961 the president of the Soviet Academy of Sciences publicly noted that high priority was being allocated to space communication within Soviet policymaking, and it was clear from his statement that television was being given priority over point-to-point communications. In April 1965, Premier Kosygin included in the same invitation as the Interkosmos proposal, that socialist states collaborate on "long-distance radio communications and television."[20] Intersputnik, then, represented one facet of the total Soviet space program, and there exists a permanent liaison committee between Interkosmos and Intersputnik.

The first draft of Intersputnik's constitution emerged in 1968 (a year in which INTELSAT was already broadcasting to sixty-three nations). A formal public announcement of the intention to set up Intersputnik was made at the UN Conference on Outer Space in Vienna in August 1969. The finished constitution was signed in Moscow by the then participating states on November 15, 1971, and came into force on July 12, 1972. It was planned to develop its operation in three stages: first, experimental use, utilizing Soviet satellites free of charge; second, offering leases on satellite communication channels; finally, a full commercial operation. By the beginning of 1977 there were ground stations in the USSR itself, at Vladimir and Dubna, and in the DDR, Poland, Czechoslovakia, Mongolia, and Cuba. There are reports in some quarters that as well as these countries formally announced as coming on line, the nations of North Korea, Libya, Angola, Mozambique, the Malagasy Republic, Sri Lanka, and some (unspecified) South American countries will be joining the system.[21] Another Soviet report cites only Libya, while the cover of the Intersputnik brochure indicates Libya, São Tomé e Príncipe, and Ethiopia, in addition to the countries officially designated as part of the system.[22] Furthermore, relations between Intersputnik and Panaftel (Pan-African Telecommunications Network) are being expanded and consolidated.[23] Not all the countries indicated are full members of Intersputnik, however: in 1984 these comprised the six eastern European nations in Comecon, plus Afghanistan, Cuba, Laos, Mongolia, North Korea, Vietnam, and South Yemen. Together with the USSR itself, this amounted to fourteen nations.[24]

The satellites used are part of the Statsionar ("stationary") network. Intersputnik uses three Gorizont ("horizon") geosynchronous satellites at 14 degrees west longitude, 53 degrees east, and 140 degrees east.[25] The only major inhabited part of the earth's surface which is not covered by them is central and western North America. Their ranges overlap at a number of points. There are plans to expand the system's technical capacity, which currently enables four to eight hours of television transmission daily,[26] another index of continuing expansion.

What information is available on its formal organizational structure, its pricing policy, its programming, and its availability? Its formal structure is laid down in the Intersputnik constitution. Intersputnik owns the equipment in space; the member countries ("recognized operating agents") own the ground stations (Article 4 of the constitution). Its ultimate authority consists of a board (Article 12), composed of one member, with one vote, from each of the signatory states. (Intersputnik is open only to states, unlike INTELSAT, which is open to nonstate subscribers.) This board meets once a year at least, generally in Moscow. The chair is rotated among the members in alphabetical order. The board elects the director-general and deputy, and must approve the structure, operation, and financial activities, of the directorate. It also sets all tariffs. Unanimity is sought, but a two-thirds vote is binding. A member stating objections in writing is not bound by a decision taken.

The permanent directorate is concerned with all administrative implementation of board decisions. Article 13 explicitly makes provision for a multinational staffing of the directorate. However, one source claims that the current directorate staff numbers only about ten technicians, after the director-general and the deputy.[27] If this is the case, it underlines the fact that programming decisions are not made by Intersputnik itself.

Intersputnik's pricing policy is probably an important component of its continuing extension. Its costs are competitive (compare the USSR's offer on its Proton rocket to Inmarsat), an important factor for many Third World economies, for which the costs of mounting a television service are extremely high. Third World countries have frequently protested high telecommunications tariffs of all kinds: witness the 1980 conference of Asian news organizations in New Delhi, which complained that satellite TV and news tariffs were actually higher for Pacific countries than for Europe.[28] The comparison of Intersputnik's rates with INTELSAT's demonstrates the overwhelming cost advantage of the former (see table 17.1).[29] With tariff differences of this kind, even INTELSAT's heavily stressed twelve price reductions from 1975 to 1981 seem to be rather beside the point.[30]

TABLE 17.1. Comparative Costs, Intelsat & Intersputnik 1981 (in Gold Francs)

Ground-station vision-channel	Intelsat	Intersputnik
1st 10 minutes	2500	990
Each following minute	78	28.2

Intersputnik's second cost advantage lies in the fact that Soviet communication satellites have very large transponders, which means the ground stations are much cheaper than INTEL-SAT's to construct.[31] The third factor, obviously, is that the USSR bears all the initial outlay of manufacturing, sending into orbit, and maintaining the satellites used. These economic factors must be weighed as seriously as the political ones when explaining the relatively rapid development of Intersputnik over recent years.

Discovering the nature of programming and the sources of decision making over programming has proved impossible at the time of writing, at least with any degree of precision. The same is true as concerns the criteria and mechanisms of program selection in the countries involved. The only regular TV program monitored in the United States at the time of writing is the nightly news and current affairs *Vremya* (time). Apart from this, only one source exists to give a "typical GORIZONT programming day" by implication from Cuba (see table 17.2).[32]

Intersputnik is now also responsible for about 40 percent of the TV transmissions by Intervidyeniye (Intervision), the Eastern European television exchange system managed by the Organization of International Radio and Television in Prague.[33] According to G. Yushkyuvichyus, deputy chair of the USSR State Committee for Television and Radio Broadcasting, Intersputnik has offered Intervidyeniye "a qualitatively new stage of development."[34] However, what these statements mean in practice is hard to specify, and the precise relationship between Intersputnik and Intervidyeniye remains to be charted.

TABLE 17.2.

13.00	UPITN news feeds transmitted in English
13.30	INTERSPUTNIK news feeds from participating nations
14.00	Panorama feed in Spanish from Czechoslovakia to Cuba
15.00	Hungarian entertainment programs from Budapest
16.00	0167 test pattern
17.00	Evening news from Moscow: Programma 1
18.00	East European entertainment programming: sporting events, dancing, music or movies
19.00	Entertainment programming from East Germany and others
21.00	Reverts to test pattern, or use by Cuba for news items, cartoons, Western entertainment, or movies

From a variety of sources, the general categories of news, politics, culture, and sport seem to be the staples of programming provision via Intersputnik. The account by Chausov, for example, describes the process of TV news exchange in the system thus:

> [TV news centers are switched into the network] offering their video news for the attention of the participants in the exchange. The working language of intercourse by the space voice communications lines is Russian, and Vremya's duty editor, sitting at the studio desk, can communicate efficiently with his colleague (*sic*) in the socialist countries.[35]

Another source indicates that the average duration of news items is about a minute, and that over 90 percent are in color. About 5 percent of items are from "liberated Vietnam, Angola, Kambodia (*sic*), Ethiopia, Mozambique, Afghanistan, and other countries." The same source also notes that Cuba receives Intervidyeniye news via Intersputnik, so it may be presumed that other Intersputnik members outside Eastern Europe also do.[36]

As regards the category "politics," this appears to refer to Party congresses and political visits. For example the twenty-sixth Soviet Party Congress in 1979 was widely broadcast through Intersputnik. So were, for example, the 1981 celebrations in Mongolia of the Mongolian People's Republic and its fifteenth Party Congress, and the sixth non-aligned nations conference in Havana in 1979.[37] Under the heading of political events might also be included such items as the televising of the award of Vietnam's Gold Star to Brezhnev, Kosygin, and Suslov in 1980,[38] and in general, visits by top dignitaries from Soviet bloc countries.

Culture in the sense of *Hochkultur* usually tends to focus on music. One obvious problem of programming is not merely cultural difference, which may be hard sometimes to translate, but the language problem itself. It is unclear how this is solved, although one Cuban source makes it plain that the Caribe Intersputnik ground station has the technology for removing the sound track from a videotape[39] which obviously would enable dubbing. In principle, the heavy stress on folkloristic dancing and similar entertainment in Eastern European programming is predictably visible on Intersputnik cultural items.

Sport, finally, is a most important element in programming, perhaps not least through being rather more telegenic than Party congresses. Soccer, football, ice hockey, even chess, are standard fare. The need for translation is also obviated. In particular, the 1980 Moscow Olympics were the occasion for a major planned expansion of Soviet TV, both nationally and internationally.[40] While they were being played, there were 460 transmissions for a total of 820 hours to over thirty countries, many via Intersputnik.[41] Enormous efforts were made in Afghanistan, Cuba, and Vietnam to ensure that ground stations came on line in time to receive the transmissions.[42]

Finally, in this description of the Intersputnik system, we should note its availability outside the nations which are members and/or users of its facilities. Clearly, any appropriately constructed dish aerial can receive it, as the Harriman Institute at Columbia University has demonstrated for eastern North America in 1984. At present, there is no regular reception by any major broadcasting agency. Indeed, the activities of some private agencies in Western European countries in organizing public viewings of Intersputnik programs have led to minor storms in local teacups. In the Netherlands in 1982, the Ministry of Culture, Recreation, and Social Welfare telephoned Amstelveen Cable TV network half an hour before a publicly advertised showing from a Gorizont satellite, and warned it that it should not proceed, as its action would be illegal. No action was taken against the network in the event. In Iceland in the same year, the People's Alliance (i.e., the Moscow-line Communist Party) celebrated the successful transmission of Intersputnik programs in an experimental showing by a private company, and were roundly taken to task for it by a conservative Icelandic newspaper, which sourly remarked on the oft-expressed hostility of the People's Alliance to having TV beamed in from the United States over military satellites.[43] However, these experimental receptions are, for the time being, just that; no agency seemed to be undertaking to relay Intersputnik television traffic on a regular basis, until the U.S. Cable News Network contract in 1985.

Pending further research into the Intersputnik system, this outline of its structure and operation is about as comprehensive a one as can be collated. We must now move to a consideration of the possible interpretations of the various forms of satellite communication cooperation and competition described above.

INTERPRETATION AND CONCLUSIONS

In any discussion of satellite competition and cooperation, it is clear that the entire communications and political context must be taken into account. No interpretation that restricts itself simply to satellites themselves can be coherent. What we have described above is set within a context of rivalry between the superpowers, expressed in communication terms in Moscow's World Service, the Voice of America, Radio Martí, Radio Free Europe, Deutsche Welle, the OIRT in Prague, and a series of other agencies. It also, as we shall see, is expressed in public relations communication of prestige projects for the admiration of the other nations on earth.

Second, "cooperation" and "competition" are not mutually exclusive spheres. A cooperative endeavor, like Interkosmos, also communicates a positive image of scientific organization and development within the Soviet bloc. The USSR's involvement with the COSPAS-SARSAT project and with Inmarsat is both a cooperative policy and a policy that indicates in principle that peace and progress are as much if not more on the Soviet agenda as on the U.S. agenda.

The politics of the first Franco-Soviet space flight indicated that this complex interrelation is much in the minds of politicians and politically engaged scientists. As we saw, Franco-Soviet cooperation in space dates back to the middle years of De Gaulle's presidency, hardly marked by an excessive enthusiasm for Communism or the Soviet Union, but dictated by a determination to pursue national independence. By the time the first joint space flight was in orbit, however, the complexion of the French government had changed, with the 1981 shift to a Socialist Party administration. As a result of the anguished squawks from across the Atlantic about the inclusion of four Communist Party ministers in the government, the internal opposition to those appointments was amplified considerably, with the result that at the time of the flight the Committee of French Physicists declared the flight "a mediocre prestige activity" and queried whether robotic flights were not a more rational policy to implement. No one above the rank of the French Moscow ambassador was present for the launch or the landing; the Centre Nationale des Etudes Spatiales was forced to say that future joint flights were not on their agenda (which was untrue) and to remind Moscow publicly that this was a scientific and not a prestige project; and upon the successful conclusion of the flight, President Mitterrand's congratulatory message included a very prominent reference to the Helsinki principles. There was even some question as to whether the astronauts would be honored by French decorations, though in the event they received the Légion d'Honneur as well as the Order of Lenin and Hero of the Soviet Union decorations. And a year later, it was announced officially that human space flights were of great scientific value, and that there would be a French astronaut in a future NASA launch.[44]

In the days of Charles de Gaulle, it is unlikely this rather inconsequential business would have been manifest. Given an administration sandwiched between its own conservative opposition and the Reagan administration, the communicative public relations aspects of the occasion were inevitable thrown into sharp relief.

There are further dimensions of Soviet collaborative space activity which require consideration in this context of interpreting cooperation and competition. Its involvement with India, as the largest non-aligned nation, is of considerable importance. Donating free access to Indian television, while a U.S.-built satellite was not functioning, was clearly useful to the Indian Ministry of Communications, but at the same time had a clear value in the competition between the United States and the Soviet Union for legitimacy in the Third World in general, and in India in particular. The remote sensing undertaken jointly between the two nations is a project whose findings may be of considerable utility to Soviet planners in developing further their already well-entrenched involvement with the Indian economy.[45] Similarly, there is a series of joint scientific space projects undertaken between Sweden and the USSR.[46] However, what are we to make of a Soviet article about remote sensing, which mostly refers to Siberia in the text, but in the first three figures displays a map of Scandinavia?[47] Given the strategic significance of Norway within NATO, the semidependent status of Finland vis-à-vis the USSR, and the repeated experiences of Sweden with Soviet submarines in the 1980s, such maps may be indicative of more than merely cooperation.

The tariff rates and structure of Intersputnik are another case in point. It is cooperative to offer lower rates to Third World nations, and yet dependence on Intersputnik's global communication links could lead to a more competitive position for the USSR in relation to its rival. The fact that INTELSAT is so much bigger than Intersputnik, and that some nations, such as Cuba and Algeria, have access to both systems, should not blind us to this dimension. As regards its structure, there has been controversy from the outset concerning its implications. The Soviet view emphasizes that "on the basis of the principles of respect for the sovereignty and independence of states, equality of rights, noninterference into internal affairs, as well as mutual help and benefit, Intersputnik should help in strengthening and developing the all-round economic, scientific-technical, cultural and other relations among its members."[48] West German and U.S. views, however, have queried whether the constitutional multinationality of the board and the directorate actually amount to very much in practice.[49]

Von Kries, for example, argued that "As actually only the Soviet Union will be in a position to make satellites and rockets available, there thus arises the organization's lasting dependence on the material space policy of its ruling member."[50] Vereshchetin has attacked this view simply by citing the director-general's statutory obligation to the multinational board in the Intersputnik constitution,[51] which has no more to commend it as an argument than the aprioristic claims of Von Kries and Doyle. Only further research will be able to shed any real light on this problem, but in general the history of multinational organizations is not one that inspires optimism concerning the lack of dominance of an already dominant nation within their counsels. In terms of current debates over direct broadcast satellites and national independence these questions are of profound importance.[52]

Given these varying aspects of the cooperation-competition problem, there is still one of major importance which has not been addressed and to which I shall devote the remainder of this paper. It is the question of the capacity of Soviet television to compete with western television in the rest of the world. It is practically regarded as a platitude that Soviet bloc television is extremely dull, a view rather confirmed by Brezhnev's public attack on it in 1978; and furthermore, the comparison of Intersputnik with INTELSAT seems inevitably destined to favor the latter, given its size, its fivefold superiority in numbers of satellites, its enormous number of national subscribers. If we look at the exchange ratios between Eurovision and Intervidyeniye in 1981, we find that out of 6,410 items on offer by Eurovision, Intervidyeniye took 5,440. In reverse, Eurovision took just 415 out of 5,395 offered by Intervidyeniye,[53] Yushkyuvichyus, cited in Chausov's article, claims that this is political prejudice, but the automatic assumption in the West would be that the Intervidyeniye items were just much less telegenic.

Of course, the Soviet Union has its own attack on the quality of western television:

> What sort of broadcasts await you? An objective chronicle of international events, or misinformation enriched with open slander of real socialism? A stage concert given by real masters of culture, or a miserable film containing a dozen murders and propagandizing the cult of violence? A soccer match between popular teams, or the lurid adventures of some western movie star?[54]

Aspects of this critique are not unfamiliar in the West, but the fact remains that however culturally impoverished, the television programs do create and sustain an audience. Can Soviet television do the same? Can Intersputnik's offerings be attractive on the open market?

In order to answer the question at all, we first have to qualify what is meant by the term "open market." We have already seen that Intersputnik's rates are highly competitive, which straightaway determines the de facto possibilities of access to certain kinds of television programming. Also, when we speak of competition on the open market, we must recall that the numerically smaller affluent markets of the West and Japan are not the only markets in existence. Popular as *Dallas* may be around the world (except in Japan!), its popularity is not necessarily an index of what Third World nations will be in a position to acquire from western television as a whole.

There are five areas in which the specifically cultural competitiveness of Soviet bloc international TV needs to be judged. One is the entree to modernity. Another is the role of Soviet Middle Asia as a model for the Third World. A third is sport on television. A fourth is the presentation of space flights themselves. The fifth is the question of everyday TV fare.

The very fact of Intersputnik's existence is influential in itself. For many small nations of the Third World, the relatively cheaply provided technology, bringing them into contact with a series of other nations in and out of the Third World, represents a much-needed entrée to the neural links of the contemporary world system (and let us not forget the point-to-point communication aspects of the system as well as its television aspects). For those nations already tied to the Soviet bloc, Intersputnik represents an even more tangible gain, given that both hardware and software may be donated as part of the Soviet aid program. Witness the Laotian Minister of Propaganda, Information, and Culture:

> It is certain that the Soviet assistance in building the relay station will be regarded as a contribution to the socialist transformation and construction in Laos . . . In addition, it is regarded as a strengthening of the solidarity and friendship between Laos and the Soviet Union.[55]

Thus the practical and the symbolic utilities of the physical provision of Intersputnik are an important part of its "message."

For Soviet policymakers themselves, the role of Soviet Middle Asia as a model for Third World development is a tirelessly repeated theme, not least in the realm of the development of communications infrastructure. It is not known how widely effective this view is within influential circles in the Third World, but it would be a gross mistake to dismiss it as useless propaganda.

For example, in September 1979 a major conference took place in Tashkent, capital of Soviet Uzbekistan, sponsored by the Soviet UNESCO Commission and the Soviet Journalists' Federation, to respond—according to Soviet sources—"to the desire expressed by many developing countries to get a better idea of the Soviet Union's experience in building up and developing a socialist system of mass information media."[56] Prominent at the meeting was the late Sharaf Rashidov, alternate member of the CPSU Central Committee and First Secretary of the Uzbekistan Communist Party's central committee, flanked by eleven other members of the Uzbek central committee to give emphasis to the involvement of Uzbek leaders in the theme of the conference. Rashidov stressed the damaging character of "imperialist domination of the spiritual life of the peoples of Asia, Africa, and Latin America," and urged the importance of "national sovereignty in the sphere of information." The implication hung in the air that these issues were long solved in Uzbekistan, and that Uzbek experience could be usefully exported in this realm.

In 1983, Tashkent was once again the site of a major international gathering of this kind, this time the fourth intergovernmental council session of the International Program for the Development of Communication (set up by UNESCO to promote New World Information Order policies). Sharaf Rashidov once again opened the meeting. Grants were announced to a series of Third World nations for training purposes in journalism and television, and for hardware acquisition. The deputy foreign minister, Stukalin, spoke of the way the USSR's "new easy communication tariffs" were saving developing countries "many millions of dollars" and urged "other powers with highly developed communication systems to do the same"—a clear reference to the United States and to INTELSAT. It was also at this gathering that a new program for training fifty journalists a year from the Third World in the Soviet Union was announced, together with a major publications program for scientific and engineering textbooks for Third

World communications systems. The development of a series of television specialists trained in the Soviet Union may be a contribution toward future receptivity to Soviet communication models, or it may be much more mixed in its effects. The intention was made abundantly clear by the head of the Soviet delegation, Krasikov, who insisted

> It is of course by no mere chance that this current fourth session is being held in Tashkent. The developing countries want to acquaint themselves with our Soviet experience in resolving the problems which they face today. During the years of Soviet power, Uzbekistan has achieved huge success in creating its mass media and in creating communications means, and our guests want to acquaint themselves with these achievements.[57]

Perhaps the single largest omission in consideration of Soviet television communication is sport. As noted above, the Moscow Olympics were a major galvanizing event for the expansion of TV facilities overseas (as well as internally). Reports from Cuba, Afghanistan, and Vietnam indicated feverish activity to try to get Intersputnik facilities on line in time for the Games. From Cuba:

> Precisely because we have that earth station, our people recently were able fully to enjoy the marvelous spectacle of the Moscow-80 Olympics, the first held in a socialist country . . . It was necessary to invest many hours over a long period of time to make this possible . . . technicians worked an average of seventeen hours a day during the fifteen days of the event . . . to insure that our people saw a spectacle that was exceptional for its beauty, color, and organization.

From Afghanistan:

> Our people have long wanted to one day watch the Olympic Games like other countries . . . Fortunately this desire has materialized with the friendly, sincere cooperation of our great friend and neighbor to the North.

From Vietnam, Hanoi radio noted that in the last weeks before the games, the Soviet technicians and their Vietnamese assistants had worked two shifts a day to complete the ground station, which had originally been scheduled for 1982. Sports coverage in the USSR is very popular and the mass appeal of sports coverage is a classic area of neglect by most academic analysts of communications, not least in the USSR. Thus the appeal of Soviet TV must be taken to include this aspect as an important strand in the total picture.

Soviet space coverage on television is also important. Not only are there many more space flights, but the involvement already noted of cosmonauts from other countries was a factor making Soviet bloc television of memorable interest to the various countries involved. Lt.-Col. Arnaldo Tamayo Mendez, for example, was not only the first Latin American in space, but also the first black person in space. As a child he had worked as a shoeshine boy and sold vegetables on the street. And he had been born in Guantánamo! The Vietnamese space pilot was credited with having shot down a B-52 during the Christmas 1972 bombing of Hanoi and Haiphong (though this was vigorously contested by U.S. sources). The date of that flight was picked to coincide with the ninetieth anniversary of Ho Chi Minh's birth, the fiftieth anniversary of the Communist Party of Vietnam, and the thirty-fifth of Vietnam's declaration of independence. Bertelan Farkas, the Hungarian astronaut, not only conducted a series of scientific experiments involving an anti-viral and anti-cancer protein, but also took up with him a series of Hungarian national dishes (Hungarian cuisine is second only to French in Europe). As already noted, Rakesh Sharma, the Indian cosmonaut, did experiments on yoga under weightless conditions. In other words, many aspects of these flights made them of intense interest, especially to the nations involved. As *Granma* put it in 1980: "From one end of the island to the other, the Cuban people have had the opportunity of sharing the emotions and tensions resulting from the launching of Soyuz-38, the entrance by Lt.-Col. Arnaldo Tamayo and Col. Yuri Romenko into the orbital station and the performance of a number of important experiments in space, through the clear pictures on television."[58]

The sensation of excitement, of belonging to the future, of being associated with a major world power with immense scientific prestige and engineering skill, should not be underestimated. Certainly for INTELSAT, among its own prestige achievements are listed its coverage of the moonwalk and of the Olympics.[59] Can Intersputnik be so far behind?

Of course, the acid test is whether the USSR can produce anything of the generalized appeal of *Dallas*. Elegant productions of Chekhov and other earlier masterpieces are well within its scope, but more popular entertainment is another matter. There are signs that the Soviet Union may be moving to discover some new formulas, beyond the World War II theme, which is hardly likely to have the same appeal outside the country. In 1984, an enormously popular serial called *Tass Is Authorized To State*, involving a good deal of adventure and action in a drama centered around the attempt to defend a new African republic from the depredations of the C.I.A., represented one such potential formula.[60] There is no reason why such formulas should not be developed within the basic boundaries of official ideology, and if they come to be popular on Soviet TV the chances are they may have a much wider appeal elsewhere.

Until then, the competitiveness of Soviet bloc television via Intersputnik is likely to be restricted to sport and to space spectaculars, including perhaps the first all-woman space crew. Certainly the public events of the West, like coronations and royal weddings, have more fascination than state visits by Soviet bloc dignitaries. Feudalism appears to be more telegenic than socialism. But it would be unwise to assume that Soviet TV, internally or externally, is incapable of becoming competitive in the quite near future.

Thus, reviewing the spectrum of Soviet space activities presented in this paper, we may deduce that the Soviets are not quite the laggards in the competition for world public opinion in this realm that they are often thought to be. Cooperation is so intensely sought after by thinking people in this heavily militarized age that an agency or nation offering it may often benefit greatly.

NOTES

1. Cited in *Air & Cosmos* (Paris), March 8, 1980, 803:35. This publication will henceforth be indicated by *A&C*.

2. K. E. Bailes, *Technology and Society Under Lenin and Stalin* (Princeton: Princeton University Press, 1978), ch. 14.

3. *National Paper USSR* (1981), p. 105; V. S. Vereshchetin, *International Cooperation in Space: Legal Questions* (Mezhdunarodnoye Sotrudnichestvo v Kosmose, Pravovyye Voprosy) (Moscow: Izdatelstvo Nauka, 1977), part 1, ch. 2.

4. *National Paper USSR* (1981), pp. 106–107.

5. *Rudé Právo*, November 16, 1978, p. 1 (Joint Publications Research Service), microfiche no. 72645 0007; henceforth only the JPRS number will be cited. *A&C*, June 28, 1980, p. 44.

6. P. Langereux, "Intercosmos 21: premier satellite russe de collecte de données," *A&C*, October 17, 1981, 877:46–47; *A&C*, April 2, 1983, 948:47.

7. P. Langereux, "Dimitru Prunariu, premier cosmonaute roumain," *A&C*, May 23, 1981, 861:47; also P. Langereux, "Pham Tuan, premier cosmonaute vietnamien," *A&C*, August 23, 1980, 822:44–46; S. Berg, "La Mission de Soyouz-39," *A&C*, April 4, 1981, 854:40, 48; *A&C*, March 12, 1983, 945:37.

8. *A&C*, September 8, 1984, 1013:63.

9. P. Langereux, "Vol spatial record de Popov et Rioumine," *A&C*, October 25, 1980, 831:59–63.

10. *National Paper USSR* (1981), p. 107; R. Chipman, ed., *The World in Space* (Englewood Cliffs, N.J.: Prentice-Hall, 1982), p. 582.

11. *National Paper USSR* (1981), p. 109.

12. *A&C*, March 22, 1980, 805:42.

13. *Ibid.*, p. 43; P. Langereux, "Projet franco-soviétique de satellite d'astronomie SIGMA," *A&C*, October 8, 1983, 970:41–42.

14. *A&C*, July 14, 1984, 997:35; *National Paper USSR* (1981), pp. 108–109.

15. *A&C*, April 2, 1983, 948:47; *Sovetskaya Rossiya*, March 12, 1983, p. 1 (Foreign Broadcast Information Service Daily Reports, Soviet Union, March 18, 1983, p. D2).

16. *A&C*, January 19, 1980, 796:35; *A&C*, January 29, 1983, 939:35; *A&C*, July 2, 1983, 961:40, 44.

17. *A&C*, December 26, 1981, 887:25; P. Langereux, "Démonstration des Sarsat-Cospas de recherche et sauvetage," *A&C*, February 12, 1983, p. 41; and *A&C*, July 21, 1984, 998:27.

18. J. N. Pelton, *INTELSAT: The Global Telecommunications Network* (Washington, D.C.: International Telecommunications Satellite Organization, 1983).

19. Committee on the Peaceful Uses of Outer Space, United Nations Document A/AC.105/335/Add.2, April 17, 1984, p. 16.

20. J. Galloway, *The Politics and Technology of Satellite Communications* (Lexington, Mass.: Heath, 1973), p. 123.

21. T. Picard, "Intersputnik: The Eastern "Brother" of INTELSAT," *Satellite Communications* (August 1982), pp. 38–44.

22. V. Romantsov and Z. Chalupsky, "The International System of Space Communication Intersputnik: Its State and Main Directions of Development," paper given at

the 33d Congress of the International Astronautical Federation, Paris, September 27–October 2, 1982; and "Intersputnik" (Moscow, n.d.).

23. R. Chipman, ed., *The World in Space.*
24. Committee on the Peaceful Uses of Outer Space, p. 16.
25. M. Long, "Moscow Around the World," *Satellite Orbit International* (September 1984), pp. 10–12.
26. Romantstov and Chalupsky, "The International System."
27. Picard, "Intersputnik."
28. *Karachi Morning News,* December 22, 1980, p. 3.
29. R. Keune, ed., *Television News in a North-South Perspective* (Bonn: Friedrich Ebert Stiftung, 1981), pp. 74–80.
30. Pelton, *INTELSAT.*
31. Bykov, *Teletranslatsia sputnikom.*
32. M. Long and J. Keating, *The World of Satellite Television* (Summertown, Tenn.: Book Publishing, 1983), p. 162.
33. V. Romantsov, "Intersputnik Orbits."
34. L. Chausov, "Intervision: A Step Into Space: How the International Exchange of Information Is Carried Out," *Pravda,* April 4, 1982, p. 4.
35. *Ibid.*
36. R. Keune, ed., *Television News,* pp. 87–90.
37. Romantsov, "Intersputnik Orbits," pp. 40–41; Montsame Radio, Ulan-Bator, May 4, 1981; and *Bohemia,* November 2, 1979.
38. Hanoi Radio, July 14, 1980.
39. Gonzalez del Pino, "Instantly," *A Juvenated Technica* (1979), 4:43.
40. V. Lebovskiy, "Television Broadcasting for the Olympics," Komsomol skaya *Pravda,* January 10, 1979, p. 2.
41. G. A. Gafurov, "Satellite Television Broadcasting Systems in the USSR, *Telecommunications and Radio Engineering* (May 1982), 37:45–48.
42. *ANIS* (Kabul), April 23, 1980, p. 3; *Granma,* August 9, 1980, p. 2; Hanoi Radio, July 16, 1980.
43. *De Volkskrant,* June 6, 1982, p. 7.
44. *A&C,* April 24, 1982, 904:51; P. Langereux, "Vol 'sans cérémonie' pour le premier spationaute francais," *A&C,* April 30, 1982, 905:46 and July 17, 1982, 916:61; P. Langereux, "La France s'engage dans les vols spatiaux humains," *A&C,* July 9, 1983, 962:30–31.
45. R. Horn, *Soviet-Indian Relations* (New York: Praeger, 1982); R. G. Ghidadubli, "India in the Soviet Union's Import Trade," *Economic and Political Weekly,* December 18, 1982, 51:2053–2060.
46. *National Paper USSR* (1981), pp. 110–111.
47. A. V. Ilyin, "The Application of Space Imagery to Geology and Mineral Exploration in the USSR: A Case History," *Advances in Space Research* (April 18, 1983), 3(2):21–23.
48. Vereshchetin, *International Cooperation in Space.*
49. W. Von Kries, "Intersputnik: Sozialistisches Gegenstuck zu Intelsat?" *Zeitschrift fur I Luftrecht und Weltraumrechtsfragen* (1973), 22(1):12–20; Doyle, cited in Vereshchetin.
50. Von Kries, "Intersputnik."
51. Vereshchetin, *International Cooperation.*

52. D. Webster, "Direct Broadcast Satellites: Proximity, Sovereignty, and National Identity," *Foreign Affairs* (Summer 1984), 5:1161–1174.

53. Chausov, "Intervision," p. 4.

54. A. Terekhov, "Space Television and Law," *Aviatsiya i Kosmonavtika* (1983), pp. 44–45.

55. Moscow Radio, February 8, 1982.

56. P. Lendvai, *The Bureaucracy of Truth* (Boulder, Colo.: Westview Press, 1981).

57. TASS, September 4, 1983.

58. John Downing, "The Intersputnik system and Soviet television," *Soviet Studies* (October 1985), 37(4):465–483.

59. Pelton, *INTELSAT*.

60. *Washington Post*, August 7, 1984.

E I G H T E E N

A Television Window on the Soviet Union

KEN SCHAFFER

The idea of pulling domestic Soviet television from the northern sky came to me through a series of powerful serendipities and by no deliberate design that I am aware of. The achievement—adapting and refining the several diverse technologies that, together, make this access possible—is no doubt significant. It has opened to American students and scholars a window on the very different world that is the Soviet Union.

The first such system ("Orbita Terminal"), installed by Orbita atop Columbia University's School of International Affairs in September 1984, has been host to processions of educators, government officials, and journalists who have visited and realized that, for the first time people of one Superpower could peer through the window into the living space of the other.

HOW SOVIET TV CAME TO COLUMBIA UNIVERSITY

I stumbled onto this. In June 1981, I was tired of cable and HBO. Living in an apartment on the northern edge of Rockefeller Center, I was a satellite systems designer who could not gain access to any geosynchronous satellites. All American U.S. domestic communications satellites are along the geostationary Clarke belt to the south. I searched for a satellite—any satellite—that I could monitor on my spectrum analyzer, but offices and apartments ten deep obscured my patio's access to U.S. domestic birds. References in the writings of two technology writers, Robert Cooper of the United States and Stephen Birkill of the United Kingdom, pointed me to the notion that there were, in fact, four satellites I could "see." They were, however, obscure satellites—and they were Russian!

Over the next year, my avocation became the development of the unique processors and components necessary to process and reconstruct video and audio from the Soviet Molniya satellites. Molniyas are the only communications satellites that are not in stationary equatorial orbits. Nearly a quarter of the Soviet Union is north of 60 degrees latitude which is the cutoff point of visibility for geosynchronous satellites. Therefore, to get continuous coverage throughout the day, the Soviets must use four successive Molniyas moving in highly inclined polar orbits throughout a twenty-four-hour period, each carrying Moscow's TV for six hours before "handing off" to the next one.

Perhaps the most interesting challenge in creating the system to receive television from the Russian satellites was to retrieve the program audio. It is not carried via sub-carrier, as is the audio on every other of the world's communications satellites, but by pulse-width-modulated lines inside the horizontal blanking interval of the video picture itself. The reason for this is that in 1963, when Molniya was first activated, the most powerful satellites the Soviets could build and launch could not afford to sacrifice the 1 dB (signal strength measurement) that an audio sub-carrier would drain from the picture. Soviet engineers devised a means, modulating a vertical line of the video signal (similar to an old film track), to convey the program audio with no power loss. With this technique, they were able to launch Molniya, which was the world's first domestic television relay satellite and meet Moscow's priority of constructing a means of national television distribution to unite its score of republics and 125 nationalities spread over 11,000 kilometers and eleven time zones.

As I continued to improve the picture from Molniya,

I found myself starting to make sense of the repeating themes and patterns, the body-language, that was program content. I do not speak the Russian language. For the first time, I found the video medium to be so powerful a context that much of the meaning came through without my knowing the verbal language. What medium could be more perfect than television to get a sense of another people? Not just for language studies—not even for Kremlinological tea leaf reading and political meteorology—but to gain a sense of the people, to listen in on their internal dialogue.

I began seeking an educational institution to sponsor me in developing a refined version of my invention. It was clear to me that this yet crude technology could bring about a major breakthrough in education. Unfortunately, the world did not beat a path to my door. Professors in five states would listen to my proposal and react: "Why, we don't even watch American television, why would we want to watch that?!"

I sensed that many of America's scholars were older (print-oriented) folks: it is one thing to watch television, another to have been brought up and nurtured by it. It is difficult to overcome cultural boundaries, whether they be between nations or between generations. My search for an educator who agreed on the value international television might have for students was to be a solitary trek for some time to come.

Three years into my searching for a sponsor, in late 1983, Jonathan Sanders contacted me after reading an article in a media magazine about an "inventor who lives in Manhattan and watches Soviet television in his living room."

Professor Sanders was the assistant director of Columbia University's heralded W. Averell Harriman Institute for the Advanced Study of the Soviet Union. In a hotel room in Moscow three years earlier, he had had his own vision: how wonderful it would be for his students to be able to watch Soviet TV and learn not only the real spoken language, but about the people, the culture, and the internal priorities and complexities of this enigmatic country.

While doors slammed on my idea across the rest of the Ivy League, Professor Sanders was, unknown to me, a mile uptown from my office, meeting with satellite companies in an effort to realize his plan. The satellite companies with whom Sanders met raised technical objections, claiming it was impossible to access domestic Soviet television from the western hemisphere. On top of that, he was told more than once, it was impossible to get HBO from Columbia because of the university's location in Manhattan's dreaded terrestrial interference "Combat Zone."

It had to be a courageous act that, in defiance of expert opinion, Sanders proceeded to seek outside sponsorship to realize his vision. His project had already become known among institute students as "Jonathan's Folly."

One hot summer afternoon in 1984 a large group of Harriman students gathered to witness our first programming for the Molniyas. But at the start, the attempt proved futile.

We began looking through the sky, north over the George Washington Bridge, for that first Molniya at 3 P.M., under my false recollection that the Soviets kept their video transponder powered twenty-four-hours a day. My recollection turned hazy when we couldn't find it. As we shortly were to learn, the Soviets only that week had switched to a new procedure of powering up a few minutes before the 4 P.M. (E.S.T.) start of their Siberian programming day. Just as much of our audience was about to go home . . . There It Was! The Molniya picture was clearer than the picture any U.S. network gets from its own local cameras, thanks to the more up-to-date French SEACAM video system used by the Soviets.

On-screen were the Kremlin towers, the sign-on logowork and music for "Vremya," the main Soviet news program, which Professor Sanders had been seeking access to for three years. Virtually every one of the hundreds of visitors who have subsequently lined up before that prototype Orbita Terminal at Columbia, including educators, journalists, and diplomats, has walked away, hours later, with his or her imagination ignited. One can see how the younger generation, who will take over the reins of our planet, might, through such technologies come to see that, balanced on a precarious hair trigger, we all share a common fate.

It is hard to hate a country when you get to know its weatherlady. I have learned this merely by watching, most frequently out of the corner of my eye, as I worked. This satellite window endows us with broader perspectives. It breaks down ethnocentric views which are now anachronisms. The view through this window goes right to the spine. The medium's most vivid "message" bypasses reason and cognition: it is that people are people.

We can have in our hands technologies that allow us to bypass our own ingrained prejudice in a way even Einstein, a poet who understood relativity in many things, would have adored. The wealth of information we should share, and mutual visibility and vulnerability we must share, make it a survival imperative that we try to understand distant worlds and ourselves better. The implications of this new technology are implicit. This powerful tool is galvanizing, and its most likely result, encouraging.

At the present time, more than sixty universities have visited Columbia and have begun to consider how this unanticipated new access can be used educationally. An encouraging number of those schools are well along on fund-raising programs to retain Orbita to implement for them its now expanded version of the terminal. While emphasis is immediately being placed on language studies, the programs will be watched by students and scholars in political science, linguistics, science, and sociology.

The University of Virginia will be the first school to have a terminal that will access not only domestic Soviet television via Molniya, but the television of half a dozen South American and European countries, too. At first, as at Columbia, programs will be available only in common areas, but soon after installation in the summer of 1986, the signal will be converted and inserted into the university's cable system in order to make the programming available, random-access mode, in the students' living environment. Such a modality promises to make the difference between traditional language "instruction" and broad language/cultural "acquisition."

What's it like to be the "inventor" of this East-West transmitter? Watching as much Soviet television as I have this year has made me a better person—and, paradoxically, more grateful than ever to be an American.

It has made me more tolerant of Russians being Russians. It has given me not only a better understanding of the Soviet Union, but has helped me realize as much about ourselves. Watching Soviet television has suggested to me that "it's all done with mirrors." We perceive through the filters we bring to view. I credit exposure to this strange window on the Soviet Union with the fact that, in the last year, many of my long-standing generic attitudes, prejudices, and preconscious aggressions have, through this sharing, begun to strip away.

Satellite-borne global data systems have, laudably, become a prerequisite for the operation of the banks and major institutions of our world. I urge using these same technologies to ensure the fact that we will continue to have a world to be home to those banks: making available off-peak transponder time to build, carry, and encourage a few merely personal, people-to-people events and exchanges that speak to the heart, bypassing the cognitive center (that Einstein warned would never get us anywhere in these growingly important matters) would be an extraordinarily rewarding "returning on investment." If we don't encourage such windows, we may never make it to 12 GHz.

WHAT'S NEXT?

Space scientists are now developing what are called Spacebridges. An interactive satellite teleconference, a Spacebridge is a two-way window in which people can look at each other while asking simple questions about one another's lives. This may include students in the Soviet Union and here in the United States; scientists there and here; physicians and physicians. Most recently, on May 7, 1985, there was a reunion of U.S. W.W. II veterans at the University of California, San Diego. The U.S. veterans spoke to their Moscow counterparts in Moscow during a satellite video teleconference moderated by Oberlin president Frederick Starr.

Several pioneering groups and individuals have quietly implemented these extraordinary uses of our hardware: what Wall Street calls teleconferences, these folks aptly call "Spacebridges." Seven such events have been produced to date, most frequently in conjuction with a small company called Internews, in New York.

The "US" Festival, held in California in 1983 and 1984, started it all by connecting 200,000 young Americans to counterparts in Moscow. A Spacebridge was used to demonstrate how new communications technologies can be used to bring the world closer.

Groups such as Unison (producers of the "US" Festivals), the Roosevelt Foundation, the Esalan Institute, and Internews along with Gosteleradio in Moscow have been slowly and delicately cooperating in exploration of the possibilities inherent in satellites. These are exciting challenges ahead that transcend national boundaries and provide tools that mankind, so long as it survives, can use to better understand the world.

Selected Bibliography

Alegrett, J. L. "U.S. Role in Satellite Communications II: Leadership Continues." *Satellite Communications* (March 1984), pp. 20–23.

Alegrett, J. L. "High Ideals: The Achilles' Heel of INTELSAT?" *Telecommunications* (May 1984), pp. 45–46.

Bailes, K. E. *Technology and Society Under Lenin and Stalin*. Princeton: Princeton University Press, (1978).

Baumol, W. J. "Contestable Markets: An Uprising in the Theory of Industry Structure." *American Economic Review* (1982), 72:1–15.

Boardman, "Regulation of Transnational Communication" *Michigan Yearbook of International Studies* (1985).

Chander, Romesh, Karnik, and Kivan, *Planning for Satellite Broadcasting: The Indian Instructional Television Experiment*. UNESCO series in Mass Communication, No. 78. New York: UN, 1979.

Chausov, L. "Intervision, A Step into Space: How the International Exchange of Information is Carried Out." *Pravda*, April 12, 1982, p. 4.

Chipman, R., ed. *The World in Space*. Englewood Cliffs, N.J.: Prentice-Hall, 1982.

Christol, Carl Q. *The Modern International Law of Outer Space*. Elmsford, N.Y.: Pergamon Press, 1982.

Clement-Jones, T. "Cable and Satellite TV in the UK and Europe: The Emerging Legal Issues." *Telecommunications Policy* (September 1983), pp. 204–214.

Codding, G. A. and A. M. Rutkowski. *The International Telecommunication Union in a Changing World*. Dedham, Mass: Artech House, 1982.

Courville, L., A. de Fontenay, and R. Dobell, eds. *Economic Analysis of Telecommunications: Theory and Applications.*" New York: North-Holland, 1983.

Demac, D. A. et al. *Equity in Orbit: The 1985 ITU Space WARC*. London: International Institute of Communications, 1985.

Dizard, W. P., Jr., *The Coming Information Age*, New York: Longman, 1982.

Downing, J. D. H. 1985. "The Intersputnik System and Soviet Television." *Soviet Studies* (October), 37(4):465–483.

Doyle, S. "An Analysis of the Socialist States' Proposal for Intersputnik," *Villanova Law Review* (1969), 15:1.

Fawcett, J. E. *Outer Space: New Challenges in Law and Policy.* New York: Oxford University Press, 1985.

Gafurov, G. A. "Satellite Television Broadcasting Systems in the USSR," *Telecommunications and Radio Engineering* (May 1982), 37:45–48.

Galloway, E. "The History and Development of Space Law: International Law and the United States." *Annals of Air and Space Law* (1982), 7:295.

Galloway, J. *The Politics and Technology of Satellite Communications.* Lexington, Mass: Heath, 1973.

Gerbner, G. and M. Siefert, eds. *World Communications: A Handbook.* New York: Longman, 1984.

Ghidadubli, R. G. "India in the Soviet Union's Import Trade." *Economic and Political Weekly* (December 18, 1982), 17(51):2053–2060.

González del Piño. "Instantly," *Juventud Tecnica* (1979), 4:43.

Hasper, D. "Receive-Only Earth Stations and Piracy of the Air Waves." *Notre Dame Law Review* (October 1982), 58:84–100.

Hinchman, W. R. "The Technological Environment for International Communications Law," In Edward McWhinney, ed. *The International Law of Communications,* Leyden: Sijthoff, 1971.

Horn, R. *Soviet-Indian Relations.* New York: Praeger, 1982.

Hudson, H. *When Telephones Reach the Village: The Role of Telecommunications in Rural Development,* Norwood, N.J.: Ablex, 1984.

Humphlett, P. E. *Use of the Geostationary Orbit and U.S. Participation in the 1985 World Administrative Radio Conference.* Congressional Research Service, Library of Congress, May 1985.

Ilyin, A. V. "The Application of Space Imagery to Geology and Mineral Exploration in the USSR: A Case History." *Advances in Space Research* (April 1983), 3(2):21–23.

International Telecommunication Union. Report on Telecommunication and the Peaceful Uses of Outer Space (booklet 33). Geneva: International Telecommunication Union, 1985.

Ito, Y., "Telecommunications and Industrial Policies in Japan: Recent Developments." In M. S. Snow, ed., *Telecommunications Regulation and Deregulation: An International Comparison.*

Jasentuliyana, N. and R. Chipman eds. "International Space Programs and Policies." *Proceedings of the 2d United Nations Conference on the Exploration and Peaceful Uses of Outer Space.* Vienna: Elsevier, 1984.

Keune, R., ed. *Television News in a North-South Perspective.* Bonn: Stiftung, 1981.

Ladd, D. "A Pavan for Print: Accommodating Copyright to the Tele-technologies." *Journal of the Copyright Society of USA* 29:246–263.

Lang, M. and J. Keating. *The World of Satellite Television,* Summertown, Tenn: Book Publishing, 1983.

Larkin, P. L. ed. *Task Force on International Satellite Communications.* New York: Twentieth Century Fund, 1969.

Lebovskiy, V. "Television Broadcasting for the 1980 Olympics." Komsomol skaya *Pravda*, January 10, 1979.

Leeson, K. W. *International Communications: Blueprint for Policy*. New York: North-Holland, 1984.

Lendvai, P. *The Bureaucracy of Truth*. Boulder, Colo.: Westview Press, 1981.

Levin, H. J. *The Invisible Resource: Use and Regulation of the Radio Spectrum*. Baltimore: Johns Hopkins University Press, 1971.

Levin, Harvey J. "Orbit and Spectrum Resource Strategies: Third World Demands." *Telecommunications Policy* (June 1981), 5(2):102–110.

Long, M. "Moscow Around the World." *Satellite Orbit International* (September 1984), pp. 10–13.

Loriot, F. "Propriete Intellectuelle et Droit Spatial." *Annals of Air and Space Law* (1979), 4:533–561.

Maddox, B. "Atlantic Communications: Skimming the Cream." *Connections: World Communications Report*, October 18, 1984.

Maddox, B. *Beyond Babel: New Directions in Communications*. Boston: Beacon Press, 1974.

Martin, James, *Communications Satellite Systems*. Telecom Library, 1978.

"Second United Nations Conference on the Exploration and Peaceful Uses of Outer Space," USSR National Paper United Nations, A.CONF.101/NP/30 (September 2, 1981).

Noam, E. M. *Telecommunications Regulation Today and Tomorrow*, New York: Harcourt Brace Jovanovich, 1983.

Nordenstreng, K. and H. I. Schiller. *National Sovereignty and International Communication*, Norwood, N.J.: Ablex, 1979.

Olian, I. "International Copyright and the Needs of Developing Countries: The Awakening at Stockholm and Paris." *Cornell International Law Journal* (May 1974), 7(2):81–112.

Olsson, A. H. "Copyright and New Communications Technology," *UNESCO Copyright Bulletin* (1977), 16(3):8–17.

Pardoe, G. K. C. *The Future for Space Technology*. Dover, N.H.: Frances Printer, 1984.

Pelton, J. N. *Global Talk*, Leyden: Sijthoff and Noordhoff, 1983.

Pelton, J. N. *INTELSAT: The Global Telecommunications Network*. Washington, D.C.: International Telecommunications Satellite Organization, 1983.

Picard, T. "Intersputnik: The Eastern 'Brother' of INTELSAT." *Satellite Communications* (August 1982), pp. 38–44.

Ploman, E. W. "The Whys and Wherefores of International Organizations." *Intermedia* (July 1980), 8(4):6–11.

Ploman, E. W. *Space, Earth, and Communication*. Westport: Greenwood Press, 1984.

Ploman, E. W. "Linking Broadcast Satellites in the UN and ITU." *Intermedia* (June 1977), 5:26–27.

Queeney, K. M. *Direct Broadcast Satellites and the United Nations*. Leyden: Sijthoff and Noordhoff, 1978.

Romantsov, V. "Intersputnik Orbits." *Aviatsiya i Kosmonavtika* (1981), 10:40–41.

Romantsov, V. and Z. Chalupsky. "The International System of Space Communication INTERSPUTNIK: Its State and Main Directions of Development," Paper given at the 33d Congress of the International Astronautical Federation, Paris, September 27–October 2, 1982.

Scantlebury, M. and C. P. Roberts, eds. *Reaching for Spectrum: WARC 1979, Center for Telecommunications in the Third World (CETTEM)*. San Jose, Costa Rica, 1982.

Snow, M. S. *International Commercial Satellite Communications: Economic and Political Issues of the First Decade of INTELSAT*. New York: Praeger, 1976.

Snow, M. S. "Telecommunication Deregulation in the Federal Republic of Germany." *Columbia Journal of World Business* (1983), 18(1):53–61.

Snow, M. S. ed. *Telecommunications Regulation and Deregulation: An International Comparison*.

Sterling, C. H., ed. *International Telecommunications and Information Policy* Washington, D.C.: Communications Press, 1984.

Terekhov, A. "Space Television and Law." *Aviatsiya i Kosmonavtika* (1983), 4:44–45.

Treaty on Principles Governing the Activities of States in the Exploration and Use of Outer Space, Including the Moon and Other Celestial Bodies. 18 U.S.T. 2410, T.I.A.S. No. 6347,610 U.N.T.S. 205.

U.S. Department of State. "International Telecommunications Satellite Consortium: Agreement Between the United States of America and Other Governments." Treaties and Other International Acts, Series 5646. Washington, D.C., 1964.

U.S. Department of State. "International Telecommunications Satellite Organization (INTELSAT): Agreement Between the United States of America and Other Governments, and Operating Agreement." Treaties and Other International Acts, Series 7532. Washington, D.C., 1971.

U.S. House of Representatives. Committee on Energy and Commerce, Majority Staff of the Subcommittee on Telecommunications, Consumer Protection, and Finance. "Telecommunications in Transition: The Status of Competition in the Telecommunications Industry." 97th Congress, 1st Sess. Washington, D.C., 1981.

U.S. Office of the President. "Memorandum for the Honorable Dean Burch, Chairman of the Federal Communications Commission." Washington, D.C., 1970. Mimeo.

U.S. President's Task Force on Communications Policy. "Final Report." Washington, D.C., 1968.

U.S. Senate. Committee on Commerce, Science, and Transportation. "Long-Range Goals in International Telecommunications and Information: An Outline for United States Policy." 98th Cong. 1st Sess. Washington, D.C., 1983.

Vereshchetin, V. S. *International Cooperation in Space: Legal Questions* (Mezhdunarodnoye Sotrudnichestvo v Kosmose. Pravovyye Voprosy). Moscow: Izdatelstvo Nauka, 1977.

Von Kries, W. "Intersputnik—sozialistisches Gegenstuck zu Intelsat?" *Zeitschrift fur Luftrecht und Weltraumrechtsfragen,* (1973), 22(1):12–20.

Webster, D. "Direct Broadcast Satellites: Proximity, Sovereignty, and National Identity." *Foreign Affairs* (Summer 1984), 62(5):1161–1174.

Contributors

CARL Q. CHRISTOL is professor of international law and political science at the University of Southern California. His credentials include a Ph.D. from the University of Chicago and an LL.B. from Yale Law School. His research interests are in legal and political problems associated with the oceans, outer space, the environment, human rights, and U.S. Constitutional law. Among many professional and honorary appointments, he was President of the American branch of the International Institute of Space Law and is a Member of the International Academy of Astronautics. He is the author of eight books, including *The Modern International Law of Outer Space* (1982) for which he received USC's Phi Kappa Award. He has been at USC since 1949, and has served as chairman of the Department of Political Science for six years.

DONNA DEMAC is a communications lawyer in New York City, and a member of the faculty of the Interactive Telecommunications program at New York University. She is the author of *Keeping America Uninformed: Government Secrecy in the 1980's* (1984); author of numerous articles on the application of new media technologies, and co-author of "Equity in Orbit: A Background Paper on the 1985 Space WARC"; and "1985 Space WARC: Assuring Access to Satellite Resources" (a report on the conference).

WILSON P. DIZARD is adjunct professor of international affairs at the School of Foreign Service, Georgetown University, and a research associate at the university's Center for Strategic and International Studies. A former Foreign Service Officer, he is a consultant to the Department of State on telecommunications and information issues, and has served on U.S. delegations to international conferences dealing with these subjects. He is the author of a number of books on communications, including *The Coming Information Age Television: A World View.*

JOHN DOWNING has been chairman of the Communications Department at Hunter College, City University of New York, since 1981, having previously worked at the University of Massachusetts, Amherst, and in London, England, as a sociologist. He is the author of a number of books and articles on communication, including an article on a variety of aspects of the Intersputnik system, to be published in *Soviet Studies.*

WILLIAM MacDONALD EVANS is the director of Space Policy and Plans for the Ministry of State for Science and Technology (MOSST) of the Government of Canada. He holds a B.S. degree in Electrical Engineering and an M.Sc. in Communications Engineering. He has worked for the Canadian government since 1967 in the Defense Research Telecommunications Establishment and the Communications Technology Satellite Project. In 1979 he joined MOSST, where he has moved from the development of science and technology policy proposals in telecommunications, transportation, and energy, to responsibility for long-term space policy.

JERRY FREIBAUM manages NASA's communications consulting program. He has directed satellite mobile communications experiments for commercial and public safety applications and managed economic and regulatory studies for a proposed commercial mobile satellite service. He has also been responsible for preparations for World Administrative Radio Conferences.

DOUGLAS GOLDSCHMIDT is a communications economist currently working as an independent consultant. Until recently he worked at the Academy for Educational Development as director of the Agency for International Development's Rural Satellite Program. He has written a number of papers dealing with the introduction of communications technologies into developing nations.

MARIO HIRSCH, a native of Luxembourg, holds a law degree and advanced degree in international law from Paris University and a doctorate in political science from the Foundation Nationale des Science Politiques in Paris. He has been Lecturer in Political Science at the University of Strasbourg, a Research Fellow at the Freie University in Berlin, and

Economics Editor of the Luxembourg weekly, *d'Letzeburger Land*. His articles have appeared in the *Revue Francaise de Science Politique, Europa Archiv,* and *West European Politics*. He was European director of Coronet Satellite Corporation in Luxembourg.

HEATHER E. HUDSON is associate professor in the College of Communications at the University of Texas at Austin. She has conducted studies on the role of telecommunications and development for the ITU, the World Bank, and the U.S. Agency for International Development. Dr. Hudson was a consultant to the Maitland Commission and is a member of the U.S. Space WARC Advisory Committee. She is the author of *When Telephones Reach the Village* (1984) and editor of *New Directions in Satellite Communications: Challenges for North and South* (1985), and has also written numerous articles on telecommunications and development.

ROY A. LAYTON is president of Sears Communication Co., a subsidiary of Sears Roebuck & Co. His company markets telecommunications services over the Sears corporate network, conducts seminars and provides telecommunications consulting services.

HARVEY LEVIN is Augustus B. Weller Professor of Economics and director of the Public Policy Workshop, Hofstra University. He is also senior research associate, Center for Policy Research, New York and Washington, D.C. He is the author of *Fact and Fancy in Television Regulation* (1980) and *The Invisible Resource* (1971). He has been a consultant on communications economics and policy issues to the Office of Technology Assessment of the U.S. Congress, the General Accounting Office, the Department of Justice, and the FCC. His research has most recently been supported by the National Science Foundation, Resources for the Future, Stanford University, and Georgetown University.

BRENDA MADDOX is London editor of *Connections*, a newsletter on telecommunications and communications editor of *The Economist* magazine.

WILLIAM H. MONTGOMERY is a graduate of the University of British Columbia (B.A. and LL.B) and the University of London, England (LL.M., air and maritime law). In 1961 he joined the Canadian Department of External Affairs and served abroad in trade and development positions in New Delhi, Bangkok, and in Geneva, where he specialized in trade policy with the GATT and UNCTAD. He also served in Ottawa as international legal adviser to Canada's senior committee responsible for participation in the 1973–79 Multilateral Trade Negotiations. In 1979 he became Canada's ambassador to Indonesia, and on return to Ottawa was appointed director general, international relations, of the Department of Communications. In this position he is responsible for the development and implementation of Canadian policies, programs and strategies related to communications interests at the international level. He was head of Canada's delegation to the 1985 Space Conference ORB-85.

SYLVIA OSPINA is a candidate for a LL.M. degree from the Institute of Air and Space Law, McGill University, Montreal, Canada. Her interest in international affairs led her to study law and to specialize in space law and international telecommunications. She received her J.D. from New York Law School (1984), and also holds a B.A. (sociology) from the University of California, and a M.S. (applied social sciences) from the State University of New York. Prior to attending law school she worked for several international organizations. She has lived and studied in South and North America, Europe, and the Middle East.

JOSEPH PELTON is director of strategic policy at INTELSAT. He was educated at the University of Tulsa, New York University, and Georgetown University, where he received his B.S. (physics, 1965), M.A. (international relations, 1973), and Ph.D. (political science, 1973). He has been involved in satellite applications since 1965, in various positions with North American Rockwell, NASA, George Washington University, Comsat and INTELSAT. He is the author of four books: *Global Communications Satellite Policy: INTELSAT, Politics and Functionalism* (1974); *Economic and Policy Problems in Satellite Communications* (1977), coedited with Marcellus Snow; *Global Talk* (1981); and *The INTELSAT Global Satellite System* (1984), coedited with Joel Alper.

KEN SCHAFFER is founder and president of the Orbita Technologies Corporation in New York City.

MARCELLUS SNOW is associate professor of economics at the University of Hawaii, where he has been since 1974. He holds a Ph.D. from the University of California, Berkeley, in economics, an M.A. in international relations from Johns Hopkins University, and an M.S. from M.I.T. in linguistics and computation. His research activity centers on economic aspects of telecommunications, information and the media, and on communications satellite systems, with special reference to the needs of Pacific island nations. He is the editor, with Joseph Pelton, of *Economic and Policy Problems in Satellite Communications* (New York: Praeger, 1977), and author of books on telecommunications regulation/deregulation, and on telecommunications/media policy in the Federal Republic of Germany.

CHRISTOPHER J. VIZAS, II, is a founding partner and officer of Orion Telecommunications, Ltd., a consulting firm specializing in international telecommunications project planning. Previously, he served on the staff of the U.S. House of Representatives as special assistant to the Chairman of the Information Subcommittee. He has also served as consultant to the White House, the Commerce Department, and private business on strategic responses to regulatory and technological constraints on new information services. He received his undergraduate, graduate, and legal education at Yale University.

FRED W. WEINGARTEN is program manager of the Communication and Information Technologies Program of the Office of Technology Assessment, U.S. Congress.

CHARLES WILK has been a Senior Policy analyst with the Office of Technology Assessment, U.S. Congress, since 1983, contributing to a major study on information technology R&D, and currently participating in planning a study on international telecommunications long-range policy issues. Between 1972 and 1983 at the National Telecommunications and Information Administration, Department of Commerce, Mr. Wilk had a variety of responsibilities, including evaluating alternative national policies and strategies for protecting federal government telecommunications and for formulating a policy for cryptography; studies in international telecommunications trade and nontariff trade barriers; Saudi Arabian communications; and foreign privacy laws and their potential impact on transborder data flows. He holds MSEE and MBA degrees.

Index